# PRODIGAL GENIUS

## THE LIFE OF NIKOLA TESLA

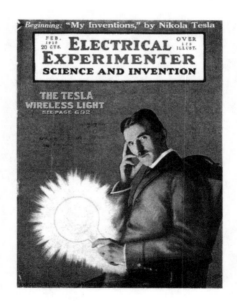

**JOHN J. O'NEILL**

Other Books in this Series:

# PRODIGAL GENIUS

## THE LIFE OF NIKOLA TESLA

John J. O'Neill

Adventures Unlimited Press

**Prodigal Genius: The Life of Nikola Tesla**

This Edition Copyright 2008
by Adventures Unlimited Press

ISBN: 978-1-931882-85-9

Published by:
Adventures Unlimited Press
One Adventure Place
Kempton, Illinois  60946  USA
auphq@frontiernet.net

www.adventuresunlimitedpress.com

10 9 8 7 6 5 4 3 2

# PRODIGAL GENIUS

## THE LIFE OF NIKOLA TESLA

WORLD WIRELESS

RETURN TO NIKOLA TESLA CO.
8 West 40 St., N.Y.

Portrait of Nikola Tesa, 1919.

# FRUITS OF GENIUS WERE SWEPT AWAY.

By a Fire the Noted Electrician, Nicola Tesla, Loses Mechanisms of Inestimable Value.

## INVENTIONS IN THE RUINS.

The Workshop Where He Evolved Ideas That Startled Electricians Entirely Destroyed.

## YEARS OF LABOR LOST.

New York Herald, March 14, 1895.

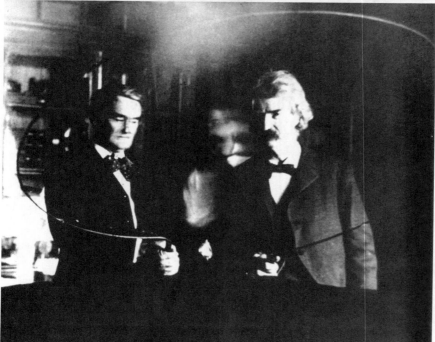

Mark Twain at Tesla's laboratory, 1895. Tesla can be seen in the bottom center.

# PRODIGAL GENIUS

# THE LIFE OF NIKOLA TESLA

Electrical oscillator activity ten million Horsepower

Power transmission without wires

New York, January 2, 1904

Tesla's letterhead showing his wireless transmission tower and other inventions.

Tesla's concept of his towers broadcasting power to electric airplanes and airships. Illustration from *Electrical Experimenter*, October 1919.

A depiction of electric anti-gravity airships that would draw power from Tesla's transmission towers. This is from *Electrical Experimenter*, October 1919.

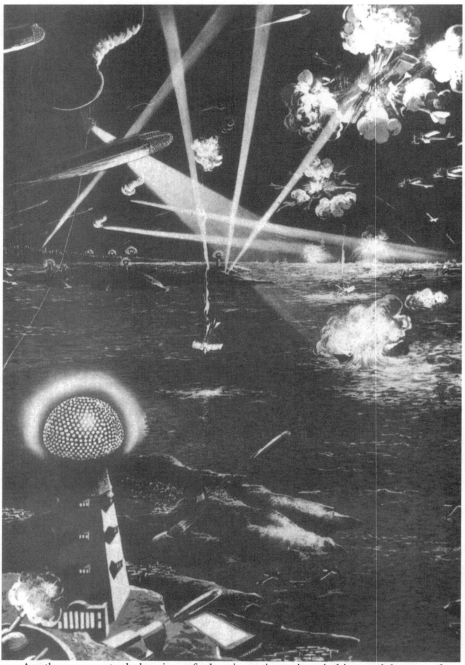

Another conceptual drawing of electric anti-gravity airships and beams of streaming power from Tesla's transmission towers. Taken from *Electrical Experimenter*, October 1919.

# CONTENTS

# FIRST PART

## LIGHT AND POWER

Nikola Tesla, 1933

# ONE

"SPECTACULAR" IS a mild word for describing the strange experiment with life that comprises the story of Nikola Tesla, and "amazing" fails to do adequate justice to the results that burst from his experiences like an exploding rocket. It is the story of the dazzling scintillations of a superman who created a new world; it is a story that condemns woman as an anchor of the flesh which retards the development of man and limits his accomplishment—and, paradoxically, proves that even the most successful life, if it does not include a woman, is a dismal failure.

Even the gods of old, in the wildest imaginings of their worshipers, never undertook such gigantic tasks of world-wide dimension as those which Tesla attempted and accomplished. On the basis of his hopes, his dreams, and his achievements he rated the status of the Olympian gods, and the Greeks would have so enshrined him. Little is the wonder that so-called practical men, with their noses stuck in profit-and-loss statements, did not understand him and thought him strange.

The light of human progress is not a dim glow that gradually becomes more luminous with time. The panorama of human evolution is illumined by sudden bursts of dazzling brilliance in intellectual accomplishments that throw their beams far ahead to give us a glimpse of the distant future, that we may more correctly guide our wavering steps today. Tesla, by virtue of the amazing discoveries and inventions which he showered on the world, becomes one of the most resplendent flashes that has ever brightened the scroll of human advancement.

Tesla created the modern era; he was unquestionably one of

the world's greatest geniuses, but he leaves no offspring, no legatees of his brilliant mind, who might aid in administering that world; he created fortunes for multitudes of others but himself died penniless, spurning wealth that might be gained from his discoveries. Even as he walked among the teeming millions of New York he became a fabled individual who seemed to belong to the far-distant future or to have come to us from the mystical realm of the gods, for he seemed to be an admixture of a Jupiter or a Thor who hurled the shafts of lightning; an Ajax who defied the Jovian bolts; a Prometheus who transmuted energy into electricity to spread over the earth; an Aurora who would light the skies as a terrestrial electric lamp; a Mazda who created a sun in a tube; a Hercules who shook the earth with his mechanical vibrators; a Mercury who bridged the ambient realms of space with his wireless waves—and a Hermes who gave birth to an electrical soul in the earth that set it pulsating from pole to pole.

This spark of intellectual incandescence, in the form of a rare creative genius, shot like a meteor into the midst of human society in the latter decades of the past century; and he lived almost until today. His name became synonymous with magic in the intellectual, scientific, engineering and social worlds, and he was recognized as an inventor and discoverer of unrivaled greatness. He made the electric current his slave. At a time when electricity was considered almost an occult force, and was looked upon with terror-stricken awe and respect, Tesla penetrated deeply into its mysteries and performed so many marvelous feats with it that, to the world, he became a master magician with an unlimited repertoire of scientific legerdemain so spectacular that it made the accomplishments of most of the inventors of his day seem like the work of toy-tinkers.

Tesla was an inventor, but he was much more than a producer of new devices: he was a discoverer of new principles, opening many new empires of knowledge which even today have been only partly explored. In a single mighty burst of invention he created the world of power of today; he brought into being our electrical power era, the rock-bottom foundation on which the industrial system of the entire world is builded; he gave us our mass-production system, for without his motors and currents it could not exist; he created the race of robots, the electrical mechanical men that are replacing human labor; he gave us every essential of modern radio; he invented the radar forty years before its use in World War II; he gave us our modern neon and other forms of gaseous-tube lighting; he gave us our fluorescent lighting; he gave us the high-frequency currents which are performing their electronic wonders throughout the industrial and medical worlds; he gave us remote control by wireless; he helped give us World War II, much against his will—for the misuse of his superpower system and his robot controls in industry made it possible for politicians to have available a tremendous surplus of power, production facilities, labor and materials, with which to indulge in the most frightful devastating war that the maniacal mind could conceive. And these discoveries are merely the inventions made by the master mind of Tesla which have thus far been utilized—scores of others remain still unused.

Yet Tesla lived and labored to bring peace to the world. He dedicated his life to lifting the burdens from the shoulders of mankind; to bringing a new era of peace, plenty and happiness to the human race. Seeing the coming of World War II, implemented and powered by his discoveries, he sought to prevent it; offered the world a device which he maintained would make any country, no matter how small, safe within its borders—and his offer was rejected.

More important by far, however, than all his stupendously significant electrical discoveries is that supreme invention—Nikola Tesla the Superman—the human instrument which shoved the world forward with an accelerating lunge like an airplane cast into the sky from a catapult. Tesla, the scientist and inventor, was himself an invention, just as much as was his alternating-current system that put the world on a superpower basis.

Tesla was a superman, a self-made superman, invented and designed specifically to perform wonders; and he achieved them in a volume far beyond the capacity of the world to absorb. His life he designed on engineering principles to enable him to serve as an automaton, with utmost efficiency, for the discovery and application of the forces of Nature to human welfare. To this end he sacrificed love and pleasure, seeking satisfaction only in his accomplishments, and limiting his body solely to serving as a tool of his technically creative mind.

With our modern craze for division of labor and specialization of effort to gain efficiency of production in our industrial machine, one hesitates to think of a future in which Tesla's invention of the superman might be applied to the entire human race, with specialization designed for every individual from birth.

The superman that Tesla designed was a scientific saint. The inventions that this scientific martyr produced were designed for the peace, happiness and security of the human race, but they have been applied to create scarcity, depressions and devastating war. Suppose the superman invention were also developed and prostituted to the purposes of war-mongering politicians? Tesla glimpsed the possibilities and suggested the community life of the bee as a threat to our social structure unless the elements of individual and community lives are properly directed and personal freedom protected.

Tesla's superman was a marvelously successful invention—for Tesla—which seemed, as far as the world could observe, to function satisfactorily. He eliminated love from his life; eliminated women even from his thoughts. He went beyond Plato, who conceived of a spiritual companionship between man and woman free from sexual desires; he eliminated even the spiritual companionship. He designed the isolated life into which no woman and no man could enter; the self-sufficient individuality from which all sex considerations were completely eliminated; the genius who would live entirely as a thinking and a working machine.

Tesla's superman invention was a producer of marvels, and he *thought* that he had, by scientific methods, succeeded in eliminating love from his life. That abnormal life makes a fascinating experiment for the consideration of the philosopher and psychologist, for he did not succeed in eliminating love. It manifested itself despite his conscientious efforts at suppression; and when it did so it came in the most fantastic form, providing a romance the like of which is not recorded in the annals of human history.

Tesla's whole life seems unreal, as if he were a fabled creature of some Olympian world. A reporter, after writing a story of his discoveries and inventions, concluded, "His accomplishments seem like the dream of an intoxicated god." It was Tesla's invention of the polyphase alternating-current system that was directly responsible for harnessing Niagara Falls and opened the modern electrical superpower era in which electricity is transported for hundred of miles, to operate the tens of thousands of mass-production factories of industrial systems. Every one of the tall Martian-like towers of the electrical transmission lines that stalk across the earth, and whose wires carry electricity to distant cities, is a monument to Tesla; every powerhouse, every dynamo and every motor that drives every machine in the country is a monument to him.

Superseding himself, he discovered the secret of transmitting electrical power to the utmost ends of the earth without wires, and demonstrated his system by which useful amounts of power could be drawn from the earth anywhere merely by making a connection to the ground; he set the entire earth in electrical vibration with a generator which spouted lightning that rivaled the fiery artillery of the heavens. It was as a minor portion of this discovery that he created the modern radio system; he planned our broadcasting methods of today, forty years ago when others saw in wireless only the dot-dash messages that might save ships in distress.

7

He produced lamps of greater brilliance and economy than those in common use today; he invented the tube, fluorescent and wireless lamps which we now consider such up-to-the-minute developments; and he essayed to set the entire atmosphere of the earth aglow with his electric currents, to change our world into a single terrestrial lamp and to make the skies at night shine as does the sun by day.

If other first-magnitude inventors and discoverers may be considered torches of progress, Tesla was a conflagration. He was the vehicle through which the blazing suns of a brighter tomorrow focused their incandescent beams on a world that was not prepared to receive their light. Nor is it remarkable that this radiant personality should have led a strange and isolated life. The value of his contributions to society cannot be overrated. We can now analyze, to some extent, the personality that produced them. He stands as a synthetic genius, a self-made superman, the greatest invention of the greatest inventor of all times. But when we consider Tesla as a human being, apart from his charming and captivating social manners, it is hard to imagine a worse nightmare than a world inhabited entirely by geniuses.

When Nature makes an experiment and achieves an improvement it is necessary that it be accomplished in such a way that the progress will not be lost with the individual but will be passed on to future generations. In man, this requires a utilization of the social values of the race, co-operation of the individual with his kind, that the improved status may be propagated and become a legacy of all. Tesla intentionally engineered love and women out of his life, and while he achieved gigantic intellectual stature, he failed to achieve its perpetuation either through his own progeny or through disciples. The superman he constructed was not great enough to embrace a wife and continue to exist as such. The love he sought to suppress in his life, and which he thought was associated only with women, is a force which, in its various aspects, links together all members of the human race.

In seeking to suppress this force entirely Tesla severed the bonds which might have brought to him the disciples who would, through other channels, have perpetuated the force of his prodigal genius. As a result, he succeeded in imparting to the world only the smallest fraction of the creative products of his synthetic superman.

The creation of a superman as demonstrated by Tesla was a grand experiment in human evolution, well worthy of the giant intellect that grew out of it, but it did not come up to Nature's standards; and the experiment will have to be made many times more before we learn how to create a super race with the minds of Teslas that can tap the hidden treasury of Nature's store of knowledge, yet endowed too with the vital power of love that will unlock forces, more powerful than any which we now glimpse, for advancing the status of the human race.

THERE was no evidence whatever that a superman was being born when the stroke of midnight between July 9 and 10, in the year 1856, brought a son, Nikola, to the home of the Rev. Milutin Tesla and Djouka, his wife, in the hamlet of Smiljan, in the Austro-Hungarian border province of Lika, now a part of Yugoslavia. The father of the new arrival, pastor of the village church, was a former student in an officers' training school who had rebelled against the restrictions of Army life and turned to the ministry as the field in which he could more satisfactorily express himself. The mother, although totally unable to read or write, was nevertheless an intellectually brilliant woman, who without the help of literal aids became really well educated.

Both father and mother contributed to the child a valuable heritage of culture developed and passed on by ancestral families that had been community leaders for many generations. The father came from a family that contributed sons in equal numbers to the Church and to the Army. The mother was a member of the Mandich family whose sons, for generations without number, had, with very few exceptions, become ministers of the Serbian Orthodox Church, and whose daughters were chosen as wives by ministers.

Djouka, the mother of Nikola Tesla (her given name in English translation would be Georgina), was the eldest daughter in a family of seven children. Her father, like her husband, was a minister of the Serbian Orthodox Church. Her mother, after a period of failing eyesight, had become blind shortly after the seventh child was born; so Djouka, the eldest daughter, at a tender age was compelled to take over the major share of her mother's duties. This not alone prevented her from attending school: her work at home so completely consumed her time that she was unable to acquire even the rudiments of reading and writing through home study. This was a strange situation in the cultured family of which she was a member. Tesla, however, always credited his unlettered mother rather than his erudite father with being the source from which he inherited his inventive ability. She devised many household labor-saving instruments. She was, in addition, a very practical individual, and her well-educated husband wisely left in her hands all business matters involving both the church and his household.

An unusually retentive memory served this remarkable woman as a good substitute for literacy. As the family moved in cultured circles she absorbed by ear much of the cultural riches of the community. She could repeat, without error or omission, thousands of verses of the national poetry of her country—the sagas of the Serbs—and could recite long passages from the Bible. She could narrate from memory the entire poetical-philosophical work *Gorski Venac* (Mountain Wreath), written by Bishop Petrovich Njegosh. She also possessed artistic talent and a versatile dexterity in her fingers for expressing it. She earned wide fame throughout the countryside for her beautiful needlework. According to Tesla, so great were her dexterity and her patience that she could, when over sixty, using only her fingers, tie three knots in an eyelash.

The remarkable abilities of this clever woman who had no formal education were transmitted to her five children. The elder son, Dane Tesla, born seven years before Nikola, was the family favorite because of the promise of an outstanding career which his youthful cleverness indicated was in store for him. He foreshadowed in his early years the strange manifestations which in his surviving brother were a prelude to greatness.

Tesla's father started his career in the military service, a likely choice for the son of an officer; but he apparently did not inherit his father's liking for Army life. So slight an incident as criticism for failure to keep his brass buttons brightly polished caused him to leave military school. He was probably more of a poet and philosopher than a soldier. He wrote poetry which was published in contemporary papers. He also wrote articles on current problems which he signed with a pseudonym, "Srbin Pravicich." This, in Serb, means "Man of Justice." He spoke, read and wrote Serbo-Croat, German and Italian. It was probably his interest in poetry and philosophy that caused him to be attracted to Djouka Mandich. She was twenty-five and Milutin was two years older. He married her in 1847. His attraction to the daughter of a pastor probably influenced his next choice of a career, for he then entered the ministry and was soon ordained a priest.

He was made pastor of the church at Senj, an important seaport with facilities for a cultural life. He gave satisfaction, but apparently he achieved success among his parishioners on the basis of a pleasing personality and an understanding of problems rather than by using any great erudition in theological and ecclesiastical matters.

A few years after he was placed in charge of this parish, a new archbishop, elevated to head of the diocese, wished to survey the capabilities of the priests in his charge and offered a prize for the best sermon preached on his official visit. The Rev. Milutin Tesla was bubbling over, at the time, with interest in labor as a major factor in social and economic problems. To preach a sermon on this topic was, from the viewpoint of expediency, a totally impractical thing to do. Nobody, however, had ever accused the Rev. Mr. Tesla of being practical, so doing the impractical thing was quite in harmony with his nature. He chose the subject which held his greatest interest; and when the archbishop arrived, he listened to a sermon on "Labor."

Months later Senj was surprised by an unanticipated visit from the archbishop, who announced that the Rev. Mr. Tesla had preached the best sermon, and awarded him a red sash which he was privileged to wear on all occasions. Shortly afterward he was made pastor at Smiljan, where his parish then embraced forty homes. He was later placed in charge of the much larger parish in the near-by city of Gospic. His first three children, Milka, Dane and Angelina, were born at Senj. Nikola and his younger sister, Marica, were born at Smiljan.

Tesla's early environment, then, was that of an agricultural community in a high plateau region near the eastern shore of the Adriatic Sea in the Velebit Mountains, a part of the Alps, a mountain chain stretching from Switzerland to Greece. He did not see his first steam locomotive until he was in his 'teens, so his aptitude for mechanical matters did not grow out of his environment.

Tesla's homeland is today called Yugoslavia, a country whose name means "Land of the Southern Slavs." It embraces several former separate countries, Serbia, Bosnia, Croatia, Montenegro, Dalmatia and also Slovenia. The Tesla and Mandich families originally came from the western part of Serbia near Montenegro. Smiljan, the village where Tesla was born, is in the province of Lika, and at the time of his birth this was a dependent province held by the Austro-Hungarian Empire as part of Croatia and Slovenia.

Tesla's surname dates back more than two and a half centuries. Before that time the family name was Draganic (pronounced as if spelled Drag'-a-nitch). The name Tesla (pronounced as spelled, with equal emphasis on both syllables), in a purely literal sense, is a trade name like Smith, Wright or Carpenter. As a common noun it describes a woodworking tool which, in English, is called an adz. This is an axe with a broad cutting blade at right angles to the handle, instead of parallel as in the more familiar form. It is used in cutting large tree trunks into squared timbers. In the Serbo-Croat language, the name of the tool is *tesla*. There is a tradition in the Draganic family that the members of one branch were given the nickname "Tesla" because of an inherited trait which caused practically all of them to have very large, broad and protruding front teeth which greatly resembled the triangular blade of the adz.

The name Draganic and derivatives of it appear frequently in other branches of the Tesla family as a given name. When used as a given name it is frequently translated "Charlotte," but as a generic term it holds the meaning "dear" and as a surname is translated "Darling."

The majority of Tesla's ancestors for whom age records are available lived well beyond the average span of life for their times, but no definite record has been found of the ancestor who, Tesla claimed, lived to be one hundred and forty years of age. (His father died at the age of fifty-nine, and his mother at seventy-one.)

Although many of Tesla's ancestors were dark eyed, his eyes were a gray-blue. He claimed his eyes were originally darker, but that as a result of the excessive use of his brain their color changed. His mother's eyes, however, were gray and so are those of some of his nephews. It is probable, therefore, that his gray eyes were inherited, rather than faded by excessive use of the brain.

Tesla grew to be very tall and very slender—tallness was a family and a national trait. When he attained full growth he was exactly two meters, or six feet two and one-quarter inches tall. While his body was slender, it was built within normal proportions. His hands, however, and particularly his thumbs, seemed unusually long.

NIKOLA's older brother Dane was a brilliant boy and his parents gloried in their good fortune in being blessed with such a fine son. There was, however, a difference of seven years in the two boys' ages, and since the elder brother died as the result of an accident at the age of twelve, when Nikola was but five years old, a fair comparison of the two seems hardly possible. The loss of their first-born son was a great blow to his mother and father; the grief and regrets of the family were manifest in idealizing his talents and predicting possibilities of greatness he might have realized, and this situation was a challenge to Nikola in his youth.

The superman Tesla developed out of the superboy Nikola. Forced to rise above the normal level by an urge to carry on for his dearly beloved departed brother, and also on his own account to exceed the great accomplishment his brother might have attained had he lived, he unconsciously drew upon strange resources within. The existence of these resources might have remained unsuspected for a lifetime, as happens with the run of individuals, if Nikola had not felt the necessity for creating a larger sphere of life for himself.

He was aware as a boy that he was not like other boys in his thoughts, in his amusements and in his hobbies. He could do the things that other lads his age usually do, and many things that they could not do. It was these latter things that interested him most, and he could find no companions who would share his enthusiasms for them. This situation caused him to isolate himself from contemporaries, and made him aware that he was destined for an unusual place if not great accomplishments in life. His boyish mind was continually exploring realms which his years had not reached, and his boyhood attainments frequently were worthy of men of mature age.

He had, of course, the usual experience of unusual incidents that fall to the lot of a small boy. One of the earliest events which Tesla recalled was a fall into a tank of hot milk that was being scalded in the process used by the natives of that region as a hygienic measure, anticipating the modern process of pasteurizing.

Shortly afterward he was accidentally locked in a remote mountain chapel which was visited only at widely separated intervals. He spent the night in the small building before his absence was discovered and his possible hiding place determined.

Living close to Nature, with ample opportunity for observing the flight of birds, which has ever filled men with envy, he did what many another boy has done with the same results. An umbrella, plus imagination, offered to him a certain solution of the problem of free flight through the air. The roof of a barn was his launching platform. The umbrella was large, but its condition was much the worse for many years of service; it turned inside out before the flight was well started. No bones were broken, but he was badly shaken up and spent the next six weeks in bed. Probably, though, he had better reason for making this experiment than most of the others who have tried it. He revealed that practically all his life he experienced a peculiar reaction when breathing deeply. When he breathed deeply he was overcome by a feeling of lightness, as if his body had lost all weight; and he should, he concluded, be able to fly through the air merely by his will to do so. He did not learn, in boyhood, that he was unusual in this respect.

One day when he was in his fifth year, one of his chums received a gift of a fishing line, and all the boys in the group planned a fishing trip. On that day he was on the outs with his chums for some unremembered reason. As a result, he was informed he could not join them. He was not permitted even to see the fishing line at close range. He had glimpsed, however, the general idea of a hook on the end of a string. In a short time he had fashioned his own interpretation of a hook. The refinement of a barb had not occurred to him and he also failed to evolve the theory of using bait when he went off on his own fishing expedition. The baitless hook failed to attract any fish

*15*

but, while dangling in the air, much to Tesla's surprise and satisfaction it snared a frog that leaped at it. He came home with a bag of nearly two dozen frogs. It may have been a day on which the fish were not biting, but at any rate his chums came home from the use of their new hook and line without any fish. His triumph was complete. When he later revealed his technique, all the boys in the neighborhood copied his hook and method, and in a short time the frog population of the region was greatly depleted.

The contents of birds' nests always excited Tesla's curiosity. He rarely disturbed their contents or occupants. On one occasion, however, he climbed a rocky crag to investigate an eagle's nest and took from it a baby eagle which he kept locked in a barn. A bird on the wing he considered fair prey for his sling shot, with which he was a star performer.

About this time he became intrigued with a piece of hollow tube cut from a cane growing in the neighborhood. This he played with until he had evolved a blow gun and later, by making a plunger and plugging one end of the tube with a wad of wet hemp, a pop gun. He then undertook the making of larger pop guns, and contrived one in which the end of the plunger was held against the chest and the tube pulled energetically toward the body. He engaged in the manufacture of this article for his chums, as a five-year-old businessman. When a number of window panes happened to get broken accidentally by getting in the way of his hemp wad, his inventive proclivities in this field were quickly curbed by the destruction of the pop guns and the administration of the parental rod.

Tesla started his formal education by attending the village school in Smiljan before he reached his fifth birthday. A few years later his father received his appointment as pastor of a church in the near-by city of Gospic, so the family moved there. This was a sad day for young Tesla. He had lived close to Nature, and loved the open country and the high mountains among which he had thus far spent all of his life. The sudden transition to the artificialities of the city was a very definite shock to him. He was out of harmony with his new surroundings.

His advent into the city life of Gospic, at the age of seven, got off to an unfortunate start. As the new minister in town, his father was anxious to have everything move smoothly. Tesla was required to dress in his best clothes and attend the Sunday services. Naturally, he dreaded this ordeal and was very happy when assigned the task of ringing the bell summoning the worshipers to the service and announcing the close of the ceremonies. This gave him an opportunity to remain unseen in the belfry while the parishioners, their daughters and dude sons were arriving and departing.

Thinking he had waited long enough after the close of the service for the church to be cleared on this first Sunday, he came downstairs three steps at a time. A wealthy woman parishioner wearing a skirt with a long train that fashionably dragged along the ground, and who had come to the service with a retinue of servants, remained after the other parishioners to have a talk with the new pastor. She was just making an impressive exit when Tesla's final jump down the stairs landed him on the train, ripping this dignity-preserving appendage from the woman's dress. Her mortification and rage and his father's anger came upon him simultaneously. Parishioners loitering outside rushed back to revel in the spectacle. Thereafter no one dared be pleasant to this youngster who had enraged the wealthy dowager who domineered it over the social community. He was practically ostracized by the parishioners, and continued so until he redeemed himself in a spectacular manner.

Tesla felt strange and defeated in his ignorance of city ways. He met the situation first by avoidance. He did not care to leave his home. The boys of his age were neatly dressed every day. There were dudes and he did not belong. Even as a child Tesla was meticulously careful in dress. At the earliest moment, however, he would slip work clothes over his dress clothes and go wandering in the woods or engage in mechanical work. He could not enjoy life if limited to the activities in which he could engage while dressed up. Tesla, however, possessed ingenuity, and there was rarely a situation in which he was not able to use it. He also possessed knowledge of the ways of Nature. These gave him a distinct superiority over the city boys.

About a year after the family moved to Gospic a new fire company was organized. It was to be supplied with a pump which would replace the useful but inadequate bucket brigade. The members of the new organization obtained brightly colored uniforms and practiced marching for parades. Eventually the new pump arrived. It was a man-power pump to be operated by sixteen men. A parade and demonstration of the new apparatus was arranged. Almost everyone in Gospic turned out for the event and followed to the river front for the pump demonstration. Tesla was among them. He paid no attention to the speeches but was all eyes for the brightly painted apparatus. He did not know how it worked but would have loved to take it apart and investigate the insides.

The time for the demonstration came when the last speaker, finishing his dedicatory address, gave the order to start the pumping operation that would send a stream of water shooting skyward from the nozzle. The eight men regimented on either side of the pump bowed and rose in alternate unison as they raised and lowered the bars that operated the pistons of the pump. But nothing else happened, not a drop of water came from the nozzle!

Officials of the fire company started feverishly to make adjustments and, after each attempt, set the sixteen men oscillating up and down at the pump handles, but each time without results. The lines of hose between the pump and the nozzle were straightened out, they were disconnected from the pump and connected again. But no water came from the far end of the hose to reward the efforts of the perspiring firemen.

Tesla was among the usual group of urchins that always manages to get inside the lines on such occasions. He tried to see everything that was going on from the closest possible vantage point and undoubtedly got on the nerves of the vexed officials when their repeated efforts were frustrated by continuous failures. As one of the officials turned for the tenth time to vent his frustration on the urchins and order them away from his range of action, Tesla grabbed him by the arm.

"I know what to do, Mister," said Tesla. "You keep pumping."

Dashing for the river, Tesla peeled his clothes off quickly and dove into the water. He swam to the suction hose that was supposed to draw the water supply from the river. He found it kinked, so that no water could flow into it, and flattened by the vacuum created by the pumping. When he straightened out the kink, the water rushed into the line. The nozzlemen had stood at their post for a long time, receiving a continuous repetition of warnings to be prepared each time an adjustment was made, but, as nothing happened on these successive occasions, they had gradually relaxed their attention and were giving little thought to the direction in which the nozzle was pointed. When the stream of water did shoot skyward, down it came on the assembled officials and townspeople. This item of unexpected drama excited the crowd at the other end of the line near the pump, and to give vent to their joy they seized the scantily dressed Tesla, boosted him to the shoulders of a couple of the firemen, and led a procession around the town. The seven-year-old Tesla was the hero of the day.

Later on Tesla, in explaining the incident, said that he had had not the faintest idea of how the pump worked; but as he watched the men struggle with it, he got an intuitive flash of knowledge that told him to go to the hose in the river. On looking back to that event, he said, he knew how Archimedes must have felt when, after discovering the law of the displacement of water by floating objects, he ran naked through the streets of Syracuse shouting "Eureka!"

AT THE age of seven Tesla had tasted the pleasures of public acclaim for his ingenuity. And further, he had done something which the dudes, the boys of his age in the city, could not do and which even their fathers could not do. He had found himself. He was now a hero, and it could be forgotten that he had jumped on a woman's skirt and ripped the train off.

Tesla never lost an opportunity to hike through the near-by mountains where he could again enjoy the pleasures of his earlier years spent so close to Nature. On these occasions he would often wonder if there was still operating a crude water wheel which he made and installed, when he was less than five years old, across the mountain brook near his home in Smiljan.

The wheel consisted of a not too well-smoothed disk cut from a tree trunk in some lumbering operations. Through its center he was able to cut a hole and force into it a somewhat straight branch of a tree, the ends of which he rested in two sticks with crotches which he forced into the rock on either bank of the brook. This arrangement permitted the lower part of the disk to dip in the water and the current caused it to rotate. To the lad there was a great deal of originality employed in making this ancient device. The wheel wobbled a bit but to him it was a marvelous piece of construction, and he got no end of pleasure out of watching his water wheel obtain power from the brook.

This experiment undoubtedly made a life-long impression on his young plastic mind and endowed him with the desire, ever afterward manifested in his work, of obtaining power from Nature's sources which are always being dissipated and always being replenished.

In this smooth-disk water wheel we find an early clue to his later invention of the smooth-disk turbine. In his later experience he discovered that all water wheels have paddles—but his little water wheel had operated without paddles.

Tesla's first experiment in original methods of power production was made when he was nine years old. It demonstrated his ingenuity and originality, if nothing else. It was a sixteen-bug-power engine. He took two thin slivers of wood, as thick as a toothpick and several times as long, and glued them together in the form of a cross, so they looked like the arms of a windmill. At the point of intersection they were glued to a spindle made of another thin sliver of wood. On this he slipped a very small pulley with about the diameter of a pea. A piece of thread acting as a driving belt was slipped over this and also around the circumference of a much larger but light pulley which was also mounted on a thin spindle. The power for this machine was furnished by sixteen May bugs (June bugs in the United States). He had collected a jar full of the insects, which were very much of a pest in the neighborhood. With a little dab of glue four bugs were affixed, heading in the same direction, to each of the four arms of the windmill arrangement. The bugs beat their wings, and if they had been free would have flown away at high speed. They were, however, attached to the cross arms, so instead they pulled them around at high speed. These, being connected by the thread belt to the large pulley, caused the latter to turn at low speed; but it developed, Tesla reports, a surprisingly large torque, or turning power.

Proud of his bug-power motor and its continous operation—the bugs did not cease flying for hours—he called in one of the boys in the neighborhood to admire it. The lad was a son of an Army officer. The visitor was amused for a short time by the bug motor, until he spied the jar of still unused May bugs. Without hesitation he opened the jar, fished out the bugs—and ate them. This so nauseated Tesla that he chased the boy out of the house and destroyed the bug motor. For years he could not tolerate the sight of May bugs without a return of this unpleasant reaction.

This event greatly annoyed Tesla because he had planned to add more spindles to the shaft and stick on more fliers until he had more than a one-hundred-bug-power motor.

# TWO

Tesla's years in school were more important for the activities in which he engaged in after-school hours than for what he learned in the classroom. At the age of ten, having finished his elementary studies in the Normal School, Tesla entered the college, called the Real Gymnasium, at Gospic. This was not an unusually early age to enter the Real Gymnasium, as that school corresponds more to our grammar school and junior high school than to our college.

One of the requirements, and one to which an unusually large percentage of the class time was devoted throughout the four years, was freehand drawing. Tesla detested the subject almost to the point of open rebellion, and his marks were accordingly very low, but not entirely owing to a lack of ability.

Tesla was left-handed as a boy, but later became ambidextrous. Left-handedness was a definite handicap in the freehand-drawing studies, but he could have done much better work than he actually produced and would have gotten higher marks if it were not for a piece of altruism in which he engaged. A student whom he could excel in drawing was striving hard for a scholarship. Were he to receive the lowest marks in freehand drawing, he would be unable to obtain the scholarship. Tesla sought to help his fellow student by intentionally getting the lowest rating in the small class.

Mathematics was his favorite subject and he distinguished himself in that study. His unusual proficiency in this field was not considered a counterbalancing virtue to make amends for his lack of enthusiasm for freehand drawing. A strange power permitted him to perform unusual feats in mathematics. He possessed it from early boyhood, but had considered it a nuisance and tried to be rid of it because it seemed beyond his control.

If he thought of an object it would appear before him exhibiting the appearance of solidity and massiveness. So greatly did these visions possess the attributes of actual objects that it was usually difficult for him to distinguish between vision and reality. This abnormal faculty functioned in a very useful fashion in his school work with mathematics.

If he was given a problem in arithmetic or algebra, it was immaterial to him whether he went to the blackboard to work it out or whether he remained in his seat. His strange faculty permitted him to see a visioned blackboard on which the problem was written, and there appeared on this blackboard all of the operations and symbols required in working out the solution. Each step appeared much more rapidly than he could work it out by hand on the actual slate. As a result, he could give the solution almost as quickly as the whole problem was stated.

His teachers, at first, had some doubts about his honesty, thinking he had worked out some clever deceit for getting the right answers. In due time their skepticism was dispelled and they accepted him as a student who was unusually apt at mental arithmetic. He would not reveal this power to anyone and would discuss it only with his mother, who in the past had encouraged him in his efforts to banish it. Now that the power had demonstrated some definite usefulness, though, he was not so anxious to be completely rid of it, but desired to bring it under his complete control.

Work that Tesla did outside school hours interested him much more than his school work. He was a rapid reader and had a memory that was retentive to the point, almost, of infallibility. He found it easy to acquire foreign languages. In addition to his native Serbo-Croat language he became proficient in the use of German, French and Italian. This opened to him great stores of knowledge to which other students did not have access, yet this knowledge, apparently, was of little use to him in his school work. He was interested in things mechanical but the school provided no manual training course. Nevertheless, he became proficient in the working of wood and metals with tools and methods of his own contriving.

In the classroom of one of the upper grades of the Real Gymnasium models of water wheels were on exhibition. They were not working models but nevertheless they aroused Tesla's enthusiasm. They recalled to him the crude wheel he had constructed in the hills of Smiljan. He had seen pictures of the magnificent Niagara Falls. Coupling the power possibilities presented by the majestic waterfalls and the intriguing possibilities he saw in the models of the water wheels, he aroused in himself a passion to accomplish a grand achievement. Waxing eloquent on the subject, he told his father, "Some day I am going to America and harness Niagara Falls to produce power." Thirty years later he was to see this prediction fulfilled.

There were many books in his father's library. The knowledge in those books interested him more than that which he received in school and he wished to spend his evenings reading them. As in other matters, he carried this to an extreme, so his father forbade him to read them, fearing that he would ruin his eyes in the poor light of tallow candles then used for illumination. Nikola sought to circumvent this ruling by taking candles to his room and reading after he was sent to bed, but his violation of orders was soon discovered and the family candle supply was hidden. Next he fashioned a candle mould out of a piece of tin and made his own candles. Then, by plugging the keyhole and the chinks around the door, he was able to spend the night hours reading volumes purloined from his father's bookshelves. Frequently, he said, he would read through the entire night and feel none the worse for the loss of sleep. Eventual discovery, however, brought paternal discipline of a vigorous nature. He was about eleven years old at this time.

Like other boys of his age he played with bows and arrows. He made bigger bows, and better, straighter shooting arrows, and his marksmanship was excellent. He was not willing to stop at that point. He started building arbalists. These could be described as bow-and-arrow guns. The bow is mounted on a frame and the string pulled back and caught on a peg from which it is released by a trigger. The arrow is laid on the midpoint of the bow, its end against the taut string. The bow lies horizontal on the frame whereas in ordinary manual shooting the bow is held in vertical position. For this reason the device is sometimes called the crossbow. In setting an arbalist the beam is placed against the abdomen and the string pulled back with all possible force. Tesla did this so often, he said, that his skin at the point of pressure became calloused until it was more like a crocodile's hide. When shot into the air the arrows from his arbalist were never recovered, for they went far out of sight. At close range they would pass through a pine board an inch thick.

Tesla got a thrill out of archery not experienced by other boys. He was, in imagination, riding those arrows which he shot out of sight into the blue vault of the heavens. That sense of exhilaration he experienced when breathing deeply gave him such a feeling of lightness he convinced himself that in this state it would be relatively easy for him to fly through the air if he only could devise some mechanical aid that would launch. him and enable him to overcome what he thought was only a slight remaining weight in his body. His earlier disastrous jump from the barn roof had not disillusioned him. His conclusions were in keeping with his sensations; but a twelve-year-old lad exploring this difficult field alone cannot be condemned too severely for not discovering that our senses sometimes deceive us, or rather that we sometimes deceive ourselves in interpreting what our senses tell us.

In breathing deeply he was overventilating his lungs, taking out some of the residual carbon dioxide which is chemical "ashes," and largely inert, and replacing it with air containing a mixture of equally inert nitrogen and very active oxygen. The latter being present in more than normal proportions immediately began to upset chemical balances throughout the body. The reaction on the brain produces a result which does not differ greatly from alcohol intoxication. A number of cults use this procedure to induce "mystical" or "occult" experiences. How was a twelve-year-old boy to know all these things? He could see that birds did an excellent job in flying. He was convinced that some day man would fly, and he wanted to produce the machine that would get him off the ground and into the air.

The big idea came to him when he learned about the vacuum —a space within a container from which all air had been exhausted. He learned that every object exposed to the air was under a pressure of about fourteen pounds per square inch, while in a vacuum objects were free of such pressure. He figured that a pressure of fourteen pounds should turn a cylinder at high speed and he could arrange to get advantage of such pressure by surrounding one half of a cylinder with a vacuum and having the remaining half of its surface exposed to air pressure. He carefully built a box of wood. At one end was an opening into which a cylinder was fitted with a very high order of accuracy, so that the box would be airtight; and on one side of the cylinder the edge of the box made a right-angle contact. On the cylinder's other side the box made a tangent, or flat, contact. This arrangement was made because he wanted the air pressure to be exerted at a tangent to the surface of the cylinder— a situation that he knew would be required in order to produce rotation. If he could get that cylinder to rotate, all he would have to do in order to fly would be to attach a propeller to a shaft from the cylinder, strap the box to his body and obtain continuous power from his vacuum box that would lift him through the air. His theory of course was fallacious, but he had no means of knowing that at the time.

The workmanship on this box was undoubtedly of a very high order, considering it was made by a self-instructed twelve-year-old mechanic. When he connected his vacuum pump, an ordinary air pump with its valves reversed, he found the box was airtight, so he pulled out all the air, watching the cylinder intently while doing so. Nothing happened for many strokes of the pump except that it made his back lame to pull the pump handle upward while he created the most "powerful" possible vacuum. He rested for a moment. He was breathing deeply from exertion, overventilating his lungs, and getting that joyous, dizzy, light-as-air feeling which was a highly satisfactory mental environment for his experiment.

Suddenly the cylinder started to turn—slowly! His experiment was a success! His vacuum-power box was working! He would fly!

Tesla was delirious with joy. He went into a state of ecstasy. There was no one with whom he could share this joy, as he had taken no one into his confidence. It was his secret and he was forced to endure its joys alone. The cylinder continued to turn slowly. It was no hallucination. It was real. It did not speed up, however, and this was disappointing. He had visualized it turning at a tremendous speed but it was actually turning extremely slowly. His idea, at least, he figured, was correct. With a little better workmanship, perhaps he could make the cylinder turn faster. He stood spellbound watching it turn at a snail's pace for less than half a minute—and then the cylinder stopped. That broke the spell and ended for the time his mental air flights.

He hunted for the trouble and quickly located what he was sure was the cause of the difficulty. Since the vacuum, he theorized, is the source of power, then, if the power stops, it must be because the vacuum is gone. His pump, he felt sure, must be leaking air. He pulled up the handle. It came up easily and that meant very definitely he had lost the vacuum in the box. He again pumped out the air—and again when he reached a high vacuum the cylinder started to turn slowly and continued to do so for a fraction of a minute. When it stopped he again pumped a vacuum and again the cylinder turned. This time he continued to operate the pump and the cylinder continued to turn. He could keep it turning as long as he desired by continuing to pump the vacuum.

There was nothing wrong with his theory, as far as he could

see. He went over the pump very carefully, making improvements which would give him a high vacuum, and studied the valve to make that a better guard of the vacuum in the box. He worked on the project for weeks but despite his best efforts he could get no better results than the slow movement of the cylinder.

Finally the truth came to him in a flash—he was losing the vacuum in the box because the air was leaking in around the cylinder on that side where the flat board was tangent to the surface of the cylinder. As the air flowed into the box it pulled the cylinder around with it very slowly. When the air stopped flowing into the box the cylinder stopped turning. He knew now his theory was wrong. He had supposed that even with the vacuum being maintained, and no air leaking in, the air pressure would be exerted at a tangent to the surface of the cylinder and the pressure would produce motion in the same way as pushing on the rim of a wheel will cause it to turn. He discovered later, however, that the air pressure is exerted at right angles to the surface of the cylinder at all points, like the direction of the spokes of a wheel, and therefore it could not be used to produce rotation in the way he planned.

This experiment, nevertheless, was not a total loss, even though it greatly disheartened him. The knowledge that the air leaking into a vacuum had actually produced even a small amount of rotation in a cylinder remained with him and led directly, many years later, to his invention of the "Tesla turbine," the steam engine that broke all records for horsepower developed per pound of weight—what he called "a power house in a hat."

NATURE seemed to be constantly engaged in staging spectacular demonstrations for young Tesla, revealing to him samples of the secret of her mighty forces.

Tesla was roaming in the mountains with some chums one winter day after a storm in which the snow fell moist and sticky.

A small snowball rolled on the ground quickly gathered more snow to itself and soon became a big one that was not too easy to move. Tiring of making snowmen and snow houses on level stretches of ground, the boys took to throwing snowballs down the sloping ground of the mountain. Most of them were duds—that is, they got stalled in the soft snow before they accumulated additional volume. A few rolled a distance, grew larger and then bogged down and stopped. One, however, found just the right conditions; it rolled until it was a large ball and then spread out, rolling up the snow at the sides as if it were rolling up a giant carpet, and then suddenly it turned into an avalanche. Soon an irresistible mass of snow was moving down the steep slope. It stripped the mountainside clean of snow, trees, soil and everything else it could carry before it and with it. The great mass landed in the valley below with a thud that shook the mountain. The boys were frightened because there was snow above them on the mountain that might have been shaken into a downward slide, carrying them along buried in it.

This event made a profound impression on Tesla and it dominated a great deal of his thinking in later life. He had witnessed a snowball weighing a few ounces starting an irresistible, devastating movement of thousands of tons of inert matter. It convinced him that there are tremendous forces locked up in Nature that can be released in gigantic amounts, for useful as well as destructive purposes, by the employment of small trigger forces. He was always on the lookout for such triggers in his later experiments.

Tesla even as a boy was an original thinker and he never hesitated to think thoughts on a grand scale, always carrying everything to its largest ultimate dimension as a means of exploring the cosmos. This is demonstrated by another event that took place the following summer. He was wandering alone in the mountains when storm clouds started to fill the sky. There was a flash of lightning and almost immediately a deluge of rain descended on him.

There was implanted in his thirteen-year-old mind on that occasion a thought which he carried with him practically all his life. He saw the lightning flash and then saw the rain come down in torrents, so he reasoned that the lightning flash produced the downpour. The idea become firmly fixed in his mind that electricity controlled the rain, and that if one could produce lightning at will, the weather would be brought under control. Then there would be no dry periods in which crops would be ruined; deserts could be turned into vineyards, the food supply of the world would be greatly increased, and there would be no lack of food anywhere on the globe. Why could he not produce lightning?

The observation and the conclusions drawn from it by young Tesla were worthy of a more mature mind, and it would require a genius among the adults to have evolved the project of controlling the world's weather through such means. There was, however, a flaw in his observation. He saw the lightning come first and the rain afterward. Further investigation would have revealed to him that the order of events was reversed higher in the air. It was the rain that came first and the lightning afterward up in the cloud. The lightning, however, arrived first because it made the trip from the cloud in less than 1/100,000 of a second, while the raindrops required several seconds to fall to the ground.

At this time there was planted in Tesla's mind the seed of a project which matured more than thirty years later when, in the mountains of Colorado, he actually produced bolts of lightning, and planned later to use them to bring rain. He never succeeded in convincing the U.S. Patent Office of the practicability of the rain-making plan.

Tesla, as a boy, knew no limits to the universe of his thinking; and as a result he built an intellectual realm sufficiently large to provide ample space in which his more mature mind could operate without encountering retarding barriers.

TESLA finished his course at the Real Gymnasium in Gospic in 1870, at the age of fourteen. He had distinguished himself as a scholar. In one grade, however, his mathematics professor gave him less than a passing mark for his year's work. Tesla felt an injustice had been done him, so he went to the director of the school and demanded that he be given the strictest kind of examination in the subject. This was done in the presence of the director and the professor, and Tesla passed it with an almost perfect mark.

His fine work at school and the recognition by the townspeople that he possessed a broader scope of knowledge than any other youth in town led the trustees of the public library to ask him to classify the books in their possession and make a catalogue. He had already read most of the books in his father's extensive library, so he was pleased to have close access to a still larger collection and undertook the task with considerable enthusiasm. He had scarcely begun work on this project when it was interrupted by a long intermittent illness. When he felt too depressed to go to the library he had quantities of the books brought to his home, and these he read while confined to his bed. His illness reached a critical stage and physicians gave up hope of saving his life.

Tesla's father knew that he was a delicate child and, having lost his other son, tried to throw every possible safeguard around this one. He was greatly pleased over his son's brilliant accomplishments in almost every activity in which he engaged, but he recognized as a danger to Nikola's health the great intensity with which he tackled projects. Nikola's trend toward engineering was to him a dangerous development, as he thought work in that field would make too heavy demands upon him, not only because of the nature of the work but in the extended years of study in which he would have to engage. If, however, the boy entered the ministry, it would not be necessary for him to extend his studies beyond the Real Gymnasium which he had just completed. For this reason his father favored a career for him in the Church.

Illness threw everything into a somber aspect. When the critical stage of his illness was reached and his strength was at its lowest ebb, Nikola manifested no inclination to help himself get better by developing an enthusiasm for anything. It was in this stage of his illness that he glanced listlessly at one of the library books. It was a volume by Mark Twain. The book held his interest and then aroused his enthusiasm for life, enabling him to pass a crisis, and his health gradually returned to normal. Tesla credited the Mark Twain book with saving his life, and when, years later, he met Twain, they became very close friends.

At the age of fifteen Tesla, in 1870, continued his studies at the Higher Real Gymnasium, corresponding to our college, at Karlovac (Carlstadt) in Croatia. His attendance at this school was made possible by an invitation from a cousin of his father's, married to a Col. Brankovic, whose home was in Karlovac, to come and live with her and her husband, a retired Army officer, while attending school. His life there was none too happy. Scarcely had he arrived when he contracted malaria from the mosquitoes in the Karlovac lowlands, and he was never free from the malady for years afterward.

Tesla relates that he was hungry all during the three years he spent at Karlovac. There was plenty of deliciously prepared food in the home, but his aunt held the theory that because his health seemed none too rugged he should not eat heavy meals. Her husband, a gruff and rugged individual, when carving a second helping for himself, would sometimes try to slip a healthy slice of meat onto Tesla's plate; but the Colonel was always overruled by his wife, who would take back the slice and carve one to the thinness of a sheet of paper, warning her husband, "Niko is delicate and we must be very careful not to overload his stomach."

His studies at Karlovac interested him, however, and he completed the four-year course in three years, tackling the school work with a dangerous enthusiasm, partly as an escape mechanism to divert his attention from the none too pleasing conditions where he was living. The lasting favorable impression which Tesla carried away from Karlovac concerned his professor of physics, a clever and original experimenter, who amazed him with the feats he performed with laboratory apparatus. He could not get enough of this course. He wanted to devote his whole time henceforth to electrical experimenting. He knew he would not be satisfied in any other field. His mind was made up; he had selected his career.

His father wrote to him shortly before his graduation advising him not to return home when school was closed but to go on a long hunting trip. Tesla, however, was anxious to get home —to surprise his parents with the good news that he had completed his work at the Higher Real Gymnasium a year ahead of schedule, and to announce his decision to make the study of electricity his life work. Greatly worried, his parents, who at that moment were making strenuous efforts to protect his health, were doubly alarmed. First, there was his violation of the instruction sent him not to return to Gospic. The reason for this advice they had not disclosed—an epidemic of cholera was raging. And second, there was his decision to enter on a career which they feared would make dangerous demands on his delicate health. On returning home, he found his plan definitely opposed. This made him very unhappy. In addition, he would shortly have to face a situation which was even more repugnant than entering upon a career in the Church, and that was the compulsory three-years' service in the Army. Those two powerful factors were operating against him and seeking to thwart him in his burning desire to start immediately unraveling the mystery and harnessing the great power of electricity.

Nothing, he thought, could exceed the difficulty of the predicament in which he found himself. In this, however, he was mistaken, for he was soon to face a much more serious problem. On the very day after his arrival home, while these issues were still red hot, he became ill with cholera. He had come home malnourished because of the inadequate amount of food to which he had been limited and the strain of his intense application to his studies. Besides, he was still suffering from malaria. Then came the cholera. Now all other problems became secondary to the immediate one of maintaining life itself against the deadly scourge. His physical condition made the doctors despair of saving him. Nevertheless, he survived the crisis, but it left him in a thoroughly weakened and run-down condition. For nine months he lay in bed almost a physical wreck. He had frequent sinking spells and from each successive one it seemed harder to rally him.

Life held no incentive for him. If he survived he would be forced to enter the Army and, if nothing happened to prevent him from finishing that term of something worse than slavery, he would be forced to study for the ministry. He did not care whether he survived or not. Left to his own decision, he would not have rallied from earlier sinking spells; but the decision was not left to him. Some force stronger than his own consciousness carried him through, but it had to succeed in spite of him and not because of any assistance he was giving. The sinking spells came on with startling regularity, each one with increasing depth. It seemed a miracle that he had come out of the last one, and now with less reserve strength he was sinking into another and edging rapidly into unconsciousness. His father entered his room and tried desperately to rouse him and stir him to a more cheerful and hopeful atitude in which he could help himself and do more than the doctors could do for him, but without results.

"I could—get well—if you—would let me—study—electrical —engineering," said the prostrate young man in a hardly audible whisper. He had scarcely enough energy left for even this effort; and having made the speech, he seemed to be dropping over the edge of nothingness. His father, bending intently over him and fearing the end had come, seized him.

"Nikola," he commanded, "you cannot go. You must stay. You will be an engineer. Do you hear me? You will go to the best engineering school in the world and you will be a great engineer. Nikola, you must come back, you must come back and become a great engineer."

The eyes of the prostrate figure opened slowly. Now there was a light shining in the eyes where before they presented a death-like glaze. The face moved a little, very little, but the slight change this movement made seemed to be in the direction of a smile. It was a smile, a weak one, and he was able to keep his eyes open although it was very apparently a struggle for him to do so.

"Thank God!" said his father. "You heard me, Nikola. You will go to an engineering school and become a great engineer. Do you understand me?"

There was not enough energy for voice but the smile became a little more definite.

Another crisis in which he had escaped death by the narrowest margin had been passed. His rise out of this situation seemed almost miraculous. It seemed to him, Tesla later related, that from that instant he felt as if he were drawing vital energy from his loved ones who surrounded him; and this he used to rally himself out of the shadow.

He was again able to whisper. "I will get well," he said weakly. He breathed deeply, as deep as his frail tired frame would permit, of the oxygen which he had found so stimulating in the past. It was the first time he had done so in the nine months since he became ill. With each breath he felt reinvigorated. He seemed to get stronger by the minute.

In a very short time he was taking nourishment and within a week he was able to sit up. In a few days more he was on his feet. Life now would be glorious. He would be an electrical engineer. Everything he dreamed of would come true. As the days passed he recovered his strength at a remarkably rapid rate and his hearty appetite returned. It was now early summer. He

*35*

would prepare himself to enter the fall term at an engineering school.

But there was something he had forgotten, everyone in the family had forgotten, in the stress of his months of illness. It was now brought sharply to his and their attention. An Army summons—he must face three years' military servitude! Was his remarkable recovery to be ruined by this catastrophe, which seemed all the worse now that his chosen career seemed other-wise nearer? Failure to respond to a military summons meant jail—and after that the service in addition. How would he solve this problem?

There is no record of what took place. This spot in his career Tesla glossed over with the statement that his father considered it advisable for him to go off on a year's hunting expedition to recover his health. At any rate, Nikola disappeared. He left with a hunting outfit and some books and paper. Where he spent the year, no one knows—probably at some hideaway in the moun-tains. In the meantime, he was a fugitive from Army service.

For any ordinary individual this situation would be a most serious one. For Tesla it had all the gravity associated with or-dinary cases, plus the complication that his family on his father's side was a traditional military family whose members had won high rank and honors in Army activities, and many of whom were now in the service of Austria-Hungary. For a member of that family to become equivalent to a "draft dodger" and a "con-scientious objector," both, was a serious blow to its prestige, and could provoke a scandal if word of the situation got into circu-lation. Tesla's father used this circumstance and the fact of Niko-la's delicate health as talking points to induce his relatives in Army positions to use their influence to enable his son to escape conscription and avoid punishment for failing to respond to the Army call. In this he was successful, apparently, but required considerable time in which to make the arrangements.

Hiding in the mountains and with a year's time to kill, on this enforced vacation Tesla was able to indulge in working out

*36*

totally fantastic plans for some gigantic projects. One of the plans was for the construction and operation of an under-ocean tube, connecting Europe and the United States, by which mail could be transported in spherical containers moved through the tube by water pressure. He discovered early in his calculations that the friction of the water on the walls of the tube would require such a tremendous amount of power to overcome it that it made the project totally impracticable. Since, however, he was working on the project entirely for his own amusement, he eliminated friction from the calculations and was then able to design a very interesting system of high-speed intercontinental mail delivery. The factor which made this interesting project impracticable—the drag of the water on the sides of the tube—Tesla was later to utilize when he invented his novel steam turbine.

The other project with which he amused himself was drawn upon an even larger scale and required a still higher order of imagination. He conceived the project of building a ring around the earth at the Equator, somewhat resembling the rings around the planet Saturn. The earth ring, however, was to be a solid structure whereas Saturn's rings are made up of dust particles.

Tesla loved to work with mathematics, and this project gave him an excellent opportunity to use all of the mathematical techniques available to him. The ring which Tesla planned was to be a rigid structure constructed on a gigantic system of scaffolding extending completely around the earth. Once the ring was complete, the scaffolding was to be removed and the ring would stay suspended in space and rotating at the same speed as the earth.

Some use might be found for the project, Tesla said, if someone could find a means of providing reactionary forces that could make the ring stand still with respect to the earth while the latter whirled underneath it at a speed of 1,000 miles per hour. This would provide a high-speed "moving" platform system of transportation which would make it possible for a person to travel around the earth in a single day.

*37*

In this project, he admitted, he encountered the same problem as did Archimedes, who said "Give me a fulcrum and a lever long enough and I will move the earth." "The fulcrum in space on which to rest the lever was no more attainable than was the reactionary force needed to halt the spinning of the hypothetical ring around the earth," said Tesla. There were a number of other factors which he found necessary to ignore in this project, but ignore them he did so that they would not interfere with his mathematical practice and his cosmical engineering plans.

With his health regained, and the danger of punishment by the Army removed, Tesla returned to his home in Gospic to remain a short time before going to Grätz, where he was to study electrical engineering as his father had promised he could do. This marked the turning point in his life. Finished with boyhood dreams and play, he was now ready to settle down to his serious life work. He had played at being a god, not hesitating to plan refashioning the earth as a planet. His life work was to produce accomplishments hardly less fantastic than his boyhood dreams.

# THREE

TESLA entered manhood with a definite knowledge that nameless forces were shaping for him an unrevealed destiny. It was a situation he had to feel rather than be able to identify and describe in words. His goal he could not see and the course leading to it he could not discern. He knew very definitely the field in which he intended to spend his life, and using such physical laws as he knew he decided to plan a life which, as an engineering project, would be operated under principles that would yield the highest index of efficiency. He did not, at this time, have a complete plan of life drawn up, but there were certain elements which he knew intuitively he would not include in his operations, so he avoided all activities and interests that would bring them in as complications. It was to be a single-purpose life, devoted entirely to science with no provisions whatever for play or romance.

It was with this philosophy of life that Tesla in 1875, at the age of 19, went to Gratz, in Austria, to study electrical engineering at the Polytechnic Institute. He intended henceforth to devote all his energies to mastering that strange, almost occult force, electricity, and to harness it for human welfare.

His first effort to put this philosophy to a practical test almost resulted in disaster despite the fact that it worked successfully. Tesla completely eliminated recreation and plunged into his studies with such enthusiastic devotion that he allowed himself only four hours' rest, not all of which he spent in slumber. He would go to bed at eleven o'clock and read himself to sleep. He was up again in the small hours of the morning, tackling his studies.

Under such a schedule he was able to pass, at the end of the first term, his examinations in nine subjects—nearly twice as many as were required. His diligence greatly impressed the members of the faculty. The dean of the technical faculty wrote to Tesla's father, "Your son is a star of first rank." The strain, however, was affecting his health. He desired to make a spectacular showing to demonstrate to his father in a practical way his appreciation of the permission he gave to study engineering. When he returned to his home at the end of the school term with the highest marks that could be awarded in all the subjects passed, he expected to be joyfully received by his father and praised for his good work. Instead, his parent showed only the slightest enthusiasm for his accomplishment but a great deal of interest in his health, and criticized Nikola for endangering it after his earlier narrow escape from death. Unknown to Tesla until several years afterward, the professor at the Polytechnic Institute had written to his father early in the term, asking him

39

to take his son out of the school, as he was in danger of killing himself through overwork.

On his return to the Institute for the second year he decided to limit his studies to physics, mechanics and mathematics. This was fortunate because it gave him more time in which to handle a situation that arose later in his studies, and was to lead to his first and perhaps greatest invention.

Early in his second year at the Institute there was received from Paris a piece of electrical equipment, a Gramme machine, that could be used as either a dynamo or motor. If turned by mechanical power it would generate electricity, and if supplied with electricity it would operate as a motor and produce mechanical power. It was a direct-current machine.

When Prof. Poeschl demonstrated the machine, Tesla was greatly impressed by its performance except in one respect—a great deal of sparking took place at the commutator. Tesla stated his objections to this defect.

"It is inherent in the nature of the machine," replied Prof. Poeschl. "It may be reduced to a great extent, but as long as we use commutators it will always be present to some degree. As long as electricity flows in one direction, and as long as a magnet has two poles each of which acts oppositely on the current, we will have to use a commutator to change, at the right moment, the direction of the current in the rotating armature."

"That is obvious," Tesla countered. "The machine is limited by the current used. I am suggesting that we get rid of the commutator entirely by using alternating current."

Long before the machine was received, Tesla had studied the theory of the dynamo and motor, and he was convinced that the whole system could be simplified in some way. The solution of the problem, however, evaded his grasp, nor was he at all sure the problem could be solved—until Prof. Poeschl gave his demonstration. The assurance then came to him like a commanding flash.

The first sources of current were batteries which produced a small steady flow. When man sought to produce electricity from mechanical power, he sought to make the same kind the batteries produced: a steady flow in one direction. The kind of current a dynamo would produce when coils of wire were whirled in a magnetic field was not this kind of current—it flowed first in one direction and then in the other. The commutator was invented as a clever device for circumventing this seeming handicap of artificial electricity and making the current come out in a one-directional flow.

The flash that came to Tesla was to let the current come out of the dynamo with its alternating directions of flow, thus eliminating the commutator, and feed this kind of current to the motors, thus eliminating the need in them for commutators. Many another scientist had played with that idea long before it occurred to Tesla, but in his case it came to him as such a vivid, illuminating flash of understanding that he knew his visualization contained the correct and practical answer. He saw both the motors and dynamos operating without commutators, and doing so very efficiently. He did not, however, see the extremely important and essential details of how this desirable result could be accomplished, but he felt an overpowering assurance that he could solve the problem. It was for this reason that he stated his objections to the Gramme machine with a great deal of confidence to his professor. What he did not expect was to draw a storm of criticism.

Prof. Poeschl, however, deviated from his set program of lectures and devoted the next one to Tesla's objections. With methodical thoroughness he picked Tesla's proposal apart and, disposing of one point after another, demonstrated its impractical nature so convincingly that he silenced even Tesla. He ended his lecture with the statement: "Mr. Tesla will accomplish great things, but he certainly never will do this. It would be equivalent to converting a steady pulling force like gravity into rotary effort. It is a perpetual motion scheme, an impossible idea."

Tesla, although silenced temporarily, was not convinced. The professor had paid him a nice compliment in devoting a whole lecture to his observation, but, as is so often the case, the compliment was loaded with what was expected by the professor to be a crushing defeat for the one whom he complimented. Tesla was nevertheless greatly impressed by his authority; and for a while he weakened in his belief that he had correctly understood his vision. It was as clear-cut and definite as the visualizations that came to him of the solutions of mathematical problems which he was always able to prove correct. But perhaps, after all, he was in this case a victim of a self-induced hallucination. All other things Prof. Poeschl taught were solidly founded on demonstrable fact, so perhaps his teacher was right in his objections to the alternating-current idea.

Deep down in his innermost being, however, Tesla held firmly to the conviction that his idea was a correct one. Criticism only temporarily submerged it, and soon it came bobbing back to the surface of his thinking. He gradually convinced himself that, contrary to his usual procedure, Prof. Poeschl had in this case demonstrated merely that he did not know how to accomplish a given result, a deficiency which he shared with everyone else in the world, and therefore could not speak with authority on this subject. And, in addition, Tesla reasoned, the closing remark with which Prof. Poeschl believed he had clinched his argument —"It would be equivalent to converting a steady pulling force like gravity into a rotary effort"—was contradicted by Nature, for was not the steady pulling force of gravity making the moon revolve around the earth and the earth revolve around the sun?

"I could not demonstrate my belief at that time," said Tesla, "but it came to me through what I might call instinct, for lack of a better name. But instinct is something which transcends knowledge. We undoubtedly have in our brains some finer fibers which enable us to perceive truths which we could not attain through logical deductions, and which it would be futile to attempt to achieve through any wilful effort of thinking."

His enthusiasm and confidence in himself restored, Tesla

tackled the problem with renewed vigor. His power of visualization—the ability to see as solid objects before him the things that he conceived in his mind, and which he had considered such a great annoyance in childhood—now proved to be of great aid to him in trying to unravel this problem. He made an elastic rebound from the intellectual trouncing administered by his professor and was tackling the problem in methodical fashion.

In his mind he constructed one machine after another, and as he visioned them before him he could trace out with his finger the various circuits through armature and field coils, and follow the course of the rapidly changing currents. But in no case did he produce the desired rotation. Practically all the remainder of the term he spent on this problem. He had passed so many examinations during the first term that he had plenty of time to spend on this problem during the second.

It seemed, however, that he was doomed to fail in this project, for at the term's end he was no nearer the solution than he was when he started. His pride had been injured and he was fighting on the defensive side. He did not know that those seeming failures in his mental and laboratory experiments were to serve later as the raw material out of which yet another vision was to be created.

A radical change had taken place in Tesla's mode of life while at Gratz. The first year he had acted like an intellectual glutton, overloading his mind and nearly wrecking his health in the process. In the second year he allowed more time for digesting the mental food of which he was partaking, and permitted himself more recreation. About this time Tesla took to card-playing as a means of relaxation. His keen mental processes and highly developed powers of deduction enabled him to win more frequently than he lost. He never retained the money he won but returned it to the losers at the end of the game. When he lost, however, this procedure was not reciprocated by the other players. He also developed a passion for billiards and chess, in both of which he became remarkably proficient.

The fondness for card-playing which Tesla developed at Gratz got him into an embarrassing situation. Toward the end of the term his father sent him money to pay for his trip to Prague and for the expenses incident to enrolling as a student at the university. Instead of going directly to Prague, Tesla returned to Gospic for a visit to the family. Sitting in at a card game with some youths of the city, Tesla found his usual luck had deserted him, and he lost the money set aside for his university expenses. He confessed to his mother what he had done. She did not criticize him. Perhaps the fates were using this method for protecting him from overwork that might ruin his health, she reasoned, since he needed rest and relaxation. Losses of money were much easier to handle than loss of health. Borrowing some money from a friend, she gave it to Tesla with the words, "Here you are. Satisfy yourself." Returning to the game, he experienced a change in luck and came out of it not only with the money his mother had given him but practically all of the university-expense money he had previously lost. These winnings he did not return to the losers as was his previous custom. He returned home, gave his mother the money she had advanced him, and announced that he would never again indulge in card-playing.

INSTEAD of going to the University of Prague in the fall of 1878 as he had planned, Tesla accepted a lucrative position that was offered him in a technical establishment at Maribor, near Gratz. He was paid sixty florins a month and a separate bonus for the completed work, a very generous compensation compared with the prevailing wages. During this year Tesla lived very modestly and saved his earnings.

The money he had saved at Maribor enabled him to pay his way through a year at the University of Prague, where he extended his studies in mathematics and physics. He continued experimenting with the one big challenging alternating-current idea that was occupying his mind. He had explored, unsuccessfully, a large number of methods and, though his failures gave support to Prof. Poeschl's contention that he would never succeed, he was unwilling to give up his theory. He still had faith that he would find the solution of his problem. He knew electrical science was young and growing, and felt deep within his consciousness that he would make the important discovery that would greatly expand the infant science to the powerful giant of the future.

It would have been a pleasure to Tesla to have continued his studies, but it now was necessary for him to make his own living. His father's death, following Tesla's graduation from the University at Prague, made it necessary for him to be self-supporting. Now he needed a job. Europe was extending an enthusiastic reception to Alexander Graham Bell's new American invention, the telephone, and Tesla heard that a central station was to be installed in Budapest. The head of the enterprise was a friend of the family. The situation seemed a promising one.

Without waiting to ascertain the situation in Budapest, Tesla, full of youthful hope and the self-assurance which is typical of the untried graduate, traveled to that city, expecting to walk into an engineering position in the new telephone project. He quickly discovered, on his arrival, that there was no position open; nor could one be created for him, as the project was still in the discussion stage.

It was, however, urgently necessary, for financial reasons, that he secure immediately a job of some kind. The best he could obtain was a much more modest one than he had anticipated. The salary was so microscopically small he would never name the amount, but it was sufficient to enable him to avoid starvation. He was employed as draftsman by the Hungarian Government in its Central Telegraph Office, which included the newly developing telephone in its jurisdiction.

It was not long before Tesla's outstanding ability attracted the attention of the Inspector in Chief. Soon he was transferred to a more responsible position in which he was engaged in designing and in making calculations and estimates in connection with new telephone installations. When the new telephone exchange was finally started in Budapest in 1881, he was placed in charge of it.

Tesla was very happy in his new position. At the age of twenty-five he was in full charge of an engineering enterprise. His inventive faculty was fully occupied and he made many improvements in telephone central-station apparatus. Here he made his first invention, then called a telephone repeater, or amplifier, but which today would be more descriptively called a loud speaker—an ancestor of the sound producer now so common in the home radio set. This invention was never patented and was never publicly described, but, Tesla later declared, in its originality, design, performance and ingenuity it would make a creditable showing alongside his better-known creations that followed. His chief interest, however, was still the alternating-current motor problem whose solution continued to elude him.

Always an indefatigable worker, always using up his available energy with the greatest number of activities he could crowd into a day, always rebelling because the days had too few hours in them and the hours too few minutes, and the seconds that composed them were of too short duration, and always holding himself down to a five-hour period of rest with only two hours of that devoted to sleep, he continually used up his vital reserves and eventually had to balance accounts with Nature. He was forced finally to discontinue work.

The peculiar malady that now affected him was never diagnosed by the doctors who attended him. It was, however, an experience that nearly cost him his life. To doctors he appeared to be at death's door. The strange manifestations he exhibited attracted the attention of a renowned physician, who declared medical science could do nothing to aid him. One of the symptoms of the illness was an acute sensitivity of all of the sense-organs. His senses had always been extremely keen, but this sensitivity was now so tremendously exaggerated that the effects were a form of torture. The ticking of a watch three rooms away sounded like the beat of hammers on an anvil. The vibration of ordinary city traffic, when transmitted through a chair or bench, pounded through his body. It was necessary to place the legs of his bed on rubber pads to eliminate the vibrations. Ordinary speech sounded like thunderous pandemonium. The slightest touch had the mental effect of a tremendous blow. A beam of sunlight shining on him produced the effect of an internal explosion. In the dark he could sense an object at a distance of a dozen feet by a peculiar creepy sensation in his forehead. His whole body was constantly wracked by twitches and tremors. His pulse, he said, would vary from a few feeble throbs per minute to more than one hundred and fifty.

Throughout this mysterious illness he was fighting with powerful desire to recover his normal condition. He had before him a task he must accomplish—he must attain the solution of the alternating-current motor problem. He felt intuitively during his months of torment that the solution was coming ever nearer, and that he must live in order to be there when it crystallized out of his unconscious mind. During this period he was unable to concentrate on this or any other subject.

Once the crisis was past and the symptoms diminished, improvement came rapidly and with it the old urge to tackle problems. He could not give up his big problem. It had become a part of him. Working on it was no longer a matter of choice. He knew that if he stopped he would die, and he knew equally well that if he failed he would perish. He was enmeshed in an invisible web of intangible structure that was tightening around him. The feeling that it was bringing the solution nearer to him—just beyond his finger tips—was cause for both regret and rejoicing. That problem when solved would leave a tremendous vacancy in his life, he feared.

Yet in spite of his feeling of optimism it was still a tremendous problem without a solution.

When the acute sensitivity reduced to normal, permitting

him to resume work, he took a walk in the city park of Budapest with a former classmate, named Szigeti, one late afternoon in February, 1882. While a glorious sunset overspread the sky with a flamboyant splash of throbbing colors, Tesla engaged in one of his favorite hobbies—reciting poetry. As a youth he had memorized many volumes, and he was now pleased to note that the terrific punishment his brain had experienced had not diminished his memory. One of the works which he could recite from beginning to end was Goethe's *Faust*.

The prismatic panorama which the sinking sun was painting in the sky reminded him of some of Goethe's beautiful lines:

> The glow retreats, done is the day of toil;
> It yonder hastes, new fields of life exploring;
> Ah, that no wing can lift me from the soil,
> Upon its track to follow, follow soaring. . . .

Tesla, tall, lean and gaunt, but with a fire in his eye that matched the flaming clouds of the heavens, waved his arms in the air and swayed his body as he voiced the undulating lines. He faced the color drama of the sky as if addressing the red-glowing orb as it flung its amorphous masses of hue, tint and chrome across the domed vault of heaven.

Suddenly the animated figure of Tesla snapped into a rigid pose as if he had fallen into a trance. Szigeti spoke to him but got no answer. Again his words were ignored. The friend was about to seize the towering motionless figure and shake him into consciousness when instead Tesla spoke.

"Watch me!" said Tesla, blurting out the words like a child bubbling over with emotion: "Watch me reverse it." He was still gazing into the sun as if that incandescent ball had thrown him into a hypnotic trance.

Szigeti recalled the image from Goethe that Tesla had been reciting: "The glow retreats . . . It yonder hastes, new fields of life exploring," a poetic description of the setting sun, and then his next words—"Watch me! Watch me reverse it." Did

*48*

Tesla mean the sun? Did he mean that he could arrest the motion of the sun about to sink below the horizon, reverse its action and start it rising again toward the zenith?

"Let us sit and rest for a while," said Szigeti. He turned him toward a bench, but Tesla was not to be moved.

"Don't you see it?" expostulated the excited Tesla. "See how smoothly it is running? Now I throw this switch—and I reverse it. See! It goes just as smoothly in the opposite direction. Watch! I stop it. I start it. There is no sparking. There is nothing on it to spark."

"But I see nothing," said Szigeti. "The sun is not sparking. Are you ill?"

"You do not understand," beamed the still excited Tesla, turning as if to bestow a benediction on his companion. "It is my alternating-current motor I am talking about. I have solved the problem. Can't you see it right here in front of me, running almost silently? It is the rotating magnetic field that does it. See how the magnetic field rotates and drags the armature around with it? Isn't it beautiful? Isn't it sublime? Isn't it simple? I have solved the problem. Now I can die happy. But I must live, I must return to work and build the motor so I can give it to the world. No more will men be slaves to hard tasks. My motor will set them free, it will do the work of the world."

Szigeti now understood. Tesla had previously told him about his attempt to solve the problem of an alternating-current motor, and he grasped the full meaning of the scientist's words. Tesla had never told him, however, about his ability to visualize objects which he conceived in his mind, so it was necessary to explain the vision he saw, and that the solution had come to him suddenly while they were admiring the sunset.

Tesla was now a little more composed, but he was floating on air in a frenzy of almost religious ecstasy. He had been breathing deeply in his excitement, and the overventilation of his lungs had produced a state of exhilaration.

Picking up a twig, he used it as a scribe to draw a diagram on

*49*

the dusty surface of the dirt walk. As he explained the technical principles of his discovery, his friend quickly grasped the beauty of his conception, and far into the night they remained together discussing its possibilities.

The conception of a rotating magnetic field was a majestically beautiful one. It introduced to the scientific world a new principle of sublime grandeur whose simplicity and utility opened a vast new empire of useful applications. In it Tesla had achieved the solution which his professor had declared was impossible of attainment.

Alternating-current motors had heretofore presented what seemed an insoluble problem because the magnetic field produced by alternating currents changed as rapidly as the current. Instead of producing a turning force they churned up useless vibration.

Up to this time everyone who tried to make an alternating-current motor used a single circuit, just as was in direct current. As a result the projected motor proved to be like a single-cylinder steam engine, stalled at dead center, at the top or bottom of the stroke.

What Tesla did was to use two circuits, each one carrying the same frequency of alternating current, but in which the current waves were out of step with each other. This was equivalent to adding to an engine a second cylinder. The pistons in the two cylinders were connected to the shaft so that their cranks were at an angle to each other which caused them to reach the top or bottom of the stroke at different times. The two could never be on dead center at the same time. If one were on dead center, the other would be off and ready to start the engine turning with a power stroke.

This analogy oversimplifies the situation, of course, for Tesla's discovery was much more far-reaching and fundamental. What Tesla had discovered was a means of creating a rotating magnetic field, a magnetic whirlwind in s ace which possessed fantastically new and intriguing properties. It was an utterly new conception. In direct-current motors a fixed magnetic field was tricked by mechanical means into producing rotation in an armature by connecting successively through a commutator each of a series of coils arranged around the circumference of a cylindrical armature. Tesla produced a field of force which rotated in space at high speed and was able to lock tightly into its embrace an armature which required no electrical connections. The rotating field possessed the property of transferring wirelessly through space, by means of its lines of force, energy to the simple closed circuit coils on the isolated armature which enabled it to build up its own magnetic field that locked itself into the rotating magnetic whirlwind produced by the field coils. The need for a commutator was completely eliminated.

Now that this magnificent solution of his most difficult scientific problem was achieved, Tesla's troubles were not over; they were just beginning; but, during the next two months, he was in a state of ecstatic pleasure playing with his new toy. It was not necessary for him to construct models of copper and iron: in his mental workshop he constructed them in wide variety. A constant stream of new ideas was continuously rushing through his mind. They came so fast, he said, that he could neither utilize nor record them all. In this short period he evolved every type of motor which was later associated with his name.

He worked out the design of dynamos, motors, transformers and all other devices for a complete alternating-current system. He multiplied the effectiveness of the two-phase system by making it operate on three or more alternating currents simultaneously. This was his famous polyphase power system.

The mental constructs were built with meticulous care as concerned size, strength, design and material; and they were tested mentally, he maintained, by having them run for weeks —after which time he would examine them thoroughly for signs of wear. Here was a most unusual mind being utilized in a most unusual way. If he at any time built a "mental machine," his memory ever afterward retained all of the details, even to the finest dimensions.

THE state of supreme happiness which Tesla was enjoying was destined soon, however, to end. The telephone central station by which he was employed, and which was controlled by Puskas, that friend of the family, was sold. When Puskas returned to Paris, he recommended Tesla for a job in the Paris establishment with which he was associated, and Tesla gladly followed up his opportunity. Paris, he reasoned, would be a wonderful springboard from which to catapult his great invention on the world.

The budding superman Tesla came to Paris light in baggage but with his head filled to bursting with his wonderful discovery of the rotating magnetic field and scores of significant inventions based on it. If he had been a typical inventor, he would have gone among people wearing a look indicating that he knew something important, but maintaining absolute secrecy concerning the nature of his inventions. He would be fearful that someone would steal his secret. But Tesla's attitude was just the reverse of this. He had something to give to the world and he wanted the world to know about it, the whole fascinating story with all the revealing technical details. He had not then learned, and never did learn, the craft of being shrewd and cunning. His life plan was on a secular basis. He cared less for the advantages of the passing moment, more for the ultimate goal; and he wanted to give his newly discovered polyphase system of alternating current to the human race that all men could benefit from it. He knew there was a fortune in his invention. How he could extract this fortune he did not know. He knew that there was a higher law of compensation under which he would derive adequate benefits from the gift to the world of his discovery. The method by which this would work out did not interest him nearly so much as the necessity for getting someone to listen to the details of his fascinating invention.

Six feet two inches tall, slender, quiet of demeanor, meticulously neat in dress, full of self-confidence, he carried himself with an air that shouted, "I defy you to show me an electrical problem I can't solve"—an attitude that was consistent with his twenty-five years, but also matched by his ability.

Through Puskas's letter of recommendation he obtained a position with the Continental Edison Company, a French company organized to make dynamos, motors and install lighting systems under the Edison patents.

He obtained quarters on the Boulevard St. Michel, but in the evenings visited and dined at the best cafés as long as his salary lasted. He made contact with many Americans engaged in electrical enterprises. Wherever he could get a patient ear, among those who had an understanding of electrical matters, he described his alternating-current system of dynamos and motors.

Did someone steal his invention? Not the slightest danger. He could not even give it away. No one was even slightly interested. The closest approach to a nibble was when Dr. Cunningham, an American, a foreman in the plant where Tesla was employed, suggested formation of a stock company.

With his great alternating-current-system invention pounding at his brain and demanding some way in which it could be developed, it was a hardship for him to be forced to work all day on direct-current machines. Nowadays, though, his health was robust. He would arise shortly after five o'clock in the morning, walk to the Seine, swim for half an hour, and then walk to Ivry, near the gates of Paris, where he was employed, a trip that required an hour of lively stepping. It was then half-past seven. The next hour he spent in eating a very substantial breakfast which never seemed sufficient to keep his appetite from developing into a disturbing factor long before noon.

The work to which he was assigned at the Continental Edison Company factory was of a variegated character, largely that of a junior engineer. In a short time he was given a traveling assignment as a "trouble shooter" which required him to visit electrical

*53*

installations in various parts of France and Germany. Tesla did not relish "trouble shooting" but he did a conscientious job and studied intensely the difficulties he encountered at each powerhouse. He was soon able to present a definite plan for improving the dynamos manufactured by his company. He presented his suggestions and received permission to apply them to some machines. When tested they were a complete success. He was then asked to design automatic regulators, for which there was a great need. These too gave an excellent performance.

The company had been placed in an embarrassing position and was threatened with heavy loss through an accident at the railroad station in Strassburg in Alsace, then in Germany, where a powerhouse and electric lights had been installed. At the opening ceremony, at which Emperor William I was present, a short circuit in the wiring caused an explosion that blew out one of the walls. The German government refused to accept the installation. Tesla was sent, early in 1883, to put the plant in working order and straighten out the situation. The technical problem presented no difficulties but he found it necessary to use a great deal of tact and good judgment in handling the mass of red tape extruded by the German government as precaution against further mishaps.

Once he got the job well under way he gave some time to constructing an actual two-phase alternating-current motor embodying his rotary-magnetic-field discovery. He had constructed so many in his mind since that never-to-be-forgotten day in Budapest when he made his great invention. He had brought materials with him from Paris for this purpose and found a machine shop near the Strassburg station where he could do some of the work. He did not have as much time available as he had expected, and, while he was a clever amateur machinist, nevertheless the work took time. He was very fussy, making every piece of metal exact in dimensions to better than the thousandth of an inch and then carefully polishing it.

Eventually there was a miscellaneous collection of parts in that

Strassburg machine shop. They had been constructed without the aid of working drawings. Tesla could project before his eyes a picture, complete in every detail, of every part of the machine. These pictures were more vivid than any blueprint and he remembered exact dimensions which he had calculated mentally for each item. He did not have to test parts through partial assembly. He knew they would fit.

From these parts Tesla quickly assembled a dynamo, to generate the two-phase alternating current which he needed to operate his alternating-current motor, and finally his new induction motor. There was no difference between the motor he built and the one which he visualized. So real was the visualized one that it had all the appearance of solidity. The one he built in the machine shop presented no elements of novelty to him. It was exactly as he had visualized it a year before. He had mentally experimented with its exact counterpart and with many variations of it during the months that had passed since the great vision came to him while rhapsodizing the sunset sky in Budapest.

The assembly completed, he started up his power generator. The time for the great final test of the validity of his theory had arrived. He would close a switch and if the motor turned his theory would be proven correct. If nothing happened, if the armature of his motor just stood still, but vibrated, his theory was not correct and he had been feeding his mind on hallucinations, based on fantasy not on fact.

He closed the switch. Instantly the armature turned, built up to full speed in a flash and then continued to operate in almost complete silence. He closed the reversing switch and the armature instantly stopped and as quickly started turning in the opposite direction. This was complete vindication of his theory.

In this experiment he had tested only his two-phase system; but he needed no laboratory demonstration to convince him that his three-phase systems for generating electricity and for using this current for transmission and power production would work even better, and that his single-phase system would work almost as well. With this working model he would now be able to convey to the minds of others the visions he had been treasuring for so long.

This test meant much more to Tesla than just the successful completion of an invention; it meant a triumph for his method of discovering new truths through the unique mental processes he used of visualizing constructs long before they were produced from materials. From these results he drew an unbounded sense of self-confidence; he could think and work his way to any goal he set.

There was good reason for Tesla's self-assurance. He had just passed his twenty-seventh birthday. It seemed to him only yesterday that Prof. Poeschl had seemingly so completely vanquished him for saying that he could operate a motor by alternating current. Now he had demonstrably accomplished what the learned professor said could never be done.

TESLA now had available a completely novel type of electrical system utilizing alternating current, which was much more flexible and vastly more efficient than the direct-current system. But now that he had it, what could he do with it? The executives of the Continental Edison Company by whom he was employed had continually refused to listen to his alternating-current theories. He felt it would be useless to try to interest them in even the working model. He had made many friends during his stay in Strassburg, among them the Mayor of the city, M. Bauzin, who shared his enthusiasm about the commercial possibilities of the new system and hoped it would result in the establishment of a new industry that would bring fame and prosperity to his city.

The Mayor brought together a number of wealthy Strassburgers. To them the new motor was shown in operation, and the new system and its possibilities described, by both Tesla and the Mayor. The demonstration was a success from the technical viewpoint but otherwise a total loss. Not one member of the group showed the slightest interest. Tesla was dejected. It was beyond his comprehension that the greatest invention in electrical science, with unlimited commercial possibilities, should be rejected so completely.

M. Bauzin assured him that he would undoubtedly receive a more satisfactory reception for his invention in Paris. Delays of officialdom in finally accepting the completed installation at the Strassburg station, however, postponed his return to Paris until the spring of 1884. Meanwhile, Tesla looked forward with pleasurable expectancy to a triumphant return to Paris. He had been promised a substantial compensation if he was successful in handling the Strassburg assignment; also, that he would be similarly compensated for the improvements in design of motors and dynamos, and for the automatic regulators for dynamos. It was possible that this would supply him with enough cash to build a full-size demonstration set for his polyphase alternating-current system, so that the tremendous advantages of his system over direct current could be shown in operation. Then he would have no trouble raising the needed capital.

When he got back to the company's offices in Paris and asked for a settlement of his Strassburg and automatic-regulator accounts, he was given what in modern terminology is called the "runaround." To use fictitious names, as Tesla told the story, the executive, Mr. Smith, who gave him the assignments, now told him he had no jurisdiction over financial arrangements; that was all in the hands of the executive, Mr. Brown. Mr. Brown explained that he administered financial matters but had no authority to initiate projects or to make payments other than those directed by the chief executive, Mr. Jones. Mr. Jones explained that such matters were in the hands of his department executives, and that he never interfered with their decisions, so Tesla must see the executive in charge of technical matters, Mr. Smith. Tesla traveled this vicious circle several times with the same result and finally gave up in disgust. He decided not to renew his offer of the alternating-current system nor to show his motor in operation, and resigned his position immediately.

Tesla was undoubtedly entitled to an amount in excess of $25,000 for the regulators he designed and for his services in Strassburg. Had the executives been endowed with even a smattering of horse sense, or the ordinary garden variety of honesty, they would have made an attempt to settle for $5,000, at the least. Tesla, hard pressed for cash, would undoubtedly have accepted such an amount, although with a feeling that he was being cheated in a large way.

Such an offer would probably have held Tesla on the payroll of the company and preserved for it the possession of the world's greatest inventor and one who at the time had definitely demonstrated he was an extremely valuable employee.

For a paltry few thousand dollars they lost not only a man who would have saved them many times that amount every year, but they also lost an opportunity to obtain world control of the greatest and most profitable electrical invention ever made.

One of the administrators of the company, Mr. Charles Batchellor, Manager of the Works, who was a former assistant and close personal friend of Thomas A. Edison, urged Tesla to go to the United States and work with Edison. There he would have a chance to work on improvements to the Edison dynamos and motors. Tesla decided to follow Mr. Batchellor's suggestion. He sold his books and all other personal possessions except a few articles which he expected to take with him. He assembled his very limited financial resources, purchased tickets for his railroad trip and transatlantic journey to New York. His baggage consisted of a small bundle of clothes carried under his arm and some other items stuffed into his pockets.

The final hours were busy ones and, as he was about to board the train, just as it was ready to pull out of the station, he discovered his package of baggage was missing. Reaching quickly for his wallet, which contained his railroad and steamship tickets and all his money, he was horrified to discover that that too was missing. There was some loose change in his pocket, how much he did not know—he did not have time to count it. His train was pulling out. What should he do? If he missed this train, he would also miss the boat—but he could not ride on either without tickets. He ran alongside the moving train, trying to make up his mind. His long legs enabled him to keep up with it without difficulty at first, but now it was gaining speed. He finally decided to jump aboard. The loose change he discovered was sufficient to take care of the railroad fare, with a negligible remainder. He explained his situation to the skeptical steamship officials and, when no one else showed up to claim his reservations on the ship up to the time of sailing, he was permitted to embark.

To one as fastidious as Tesla, a long steamship journey without adequate clothing was a trying experience. He had expected to encounter annoyances when getting along with the minimum amount of clothing which he planned carrying with him, but when even that limited layout was lost the annoyance became hardship. Coupled with this was the memory of disappointment and resentment over his recent experiences.

The ship offered little to interest him. He explored it thoroughly and in doing so made some contacts with members of the ship's company. There was unrest among the crew. There was unrest in Tesla also. He extended sympathy to members of the crew in their claimed unjust treatment. The grievances affecting the crew had built up one of those situations in which a small spark can cause a large explosion. The spark flew somewhere on the ship while Tesla was below decks in the crew's quarters. The captain and officers got tough and, with some loyal members of the crew, decided to settle the trouble with belaying pins as clubs. It quickly became a battle royal. Tesla found himself in the middle of a fight in which when anyone saw a head he hit it.

Had Tesla not been young as well as tall and strong, his useful career might have ended at this point. He had long arms in proportion to his six feet two inches of height. The fist at the end of his arm could reach as far as a club in the hands of an adversary, and his height enabled him to tower over the other fighters so his head was not easy to reach. He struck hard and often, never knowing for or against which side he was fighting. He was on his feet when the fight was over, something which could not be said of a score of the crew members. The officers had subdued what they called a mutiny, but they too carried indications that they had been through a battle. Tesla was definitely not invited to sit at the captain's table during the voyage.

He spent the remainder of his journey nursing scores of bruises and sitting in meditation at the stern of the ship, which too slowly made its way to New York. Soon he would set foot on the "land of golden promise" and meet the famous Mr. Edison. He was destined to learn that it was really a "land of golden promise" —but also to discover something that would open his eyes about the fulfillment of promises.

# FOUR

WHEN Tesla stepped out of the Immigration Office at Castle Garden, Manhattan, in the summer of 1884, his possessions consisted of four cents, a book of his own poems, a couple of technical articles he had written, calculations for designing a flying machine, and some mathematical work done in an effort to solve an extremely difficult integral. He had Mr. Batchellor's letter introducing him to Mr. Edison, and the address of a friend. In this letter to Edison, Batchellor wrote: "I know two great men and you are one of them; the other is this young man."

Lacking carfare, Tesla had to walk the several miles to his friend's home. The first person he spoke to, seeking traveling directions, was a policeman, a gruff individual. The way he

supplied the information suggested to Tesla that he was willing
to start a fight on the subject. Although Tesla spoke English very
well, all he understood of the policeman's lingo was the direction
in which he pointed his club.

While walking in what he believed was the right direction,
wondering how he would be able to contrive a meal and lodgings out of four cents should he be unable to locate his friend,
he passed a shop in which he could see a man working on an
electrical machine that seemed to him familiar. He entered just
as the man was about to give up as impossible the task of repair-
ing the device.

"Let me do it," said Tesla, "I will make it operate." And without more ado he tackled the job. It proved to be a difficult task
but eventually the machine was working again.

"I need a man like you to handle these blankety-blank foreign
machines," said the man. "Do you want a job?"

Tesla thanked him and told him he was on his way to another
job, whereupon the man handed him twenty dollars. Tesla had
expected no compensation for doing what he considered a slight
favor, and said so, but the man insisted his work was worth that
much, and he was glad to pay it. Never was Tesla more thankful
for a windfall. He was now assured of food and lodgings for the
time being. With the aid of walking directions, this time more
graciously given, he located his friend and was a guest at his
home overnight. The next day he went to Edison's New York
headquarters, then on South Fifth Avenue (now West Broadway).

The introduction by Mr. Batchellor gave him ready access to
Mr. Edison, who was busily engaged in problems in connection
with his new generating station and electric-light system—the
former located in downtown Pearl Street and serving a relatively
small radius of territory.

Tesla was favorably impressed by Edison on their first meeting. He marveled that a man so limited in education could accomplish so much in so technical a field as electricity. It caused

Tesla to wonder if all the time he had spent in gaining an education of very broad scope had not been wasted. Would he have
been further ahead if he had started his practical work on the
basis of experience, as Edison had done? He definitely decided,
however, before many days had passed, that the time and effort
he had spent on his education constituted the wisest kind of an
investment.

Edison, for his part, was none too favorably impressed by
Tesla. Edison was an inventor who got his results by trial-and-
error methods. Tesla calculated everything mentally and solved
his problems before doing any "work" on them. As a result,
the two great men spoke an entirely different technical language.
There was one more very important difference. Edison belonged
to the direct-current and Tesla to the alternating-current school
of thought. The electricians of that day could, and did, become
highly emotional over their differences of opinion on this subject.
Discussions roused all the fervor of a religious or political debate,
and everything unpleasant was associated with the adherents on
the other side of the discussion. The least unpleasant thought
applied to an opponent was that he was of a low order of mental-
ity. When Tesla enthusiastically described his polyphase system
and told Edison he believed alternating current was the only
practical kind of current to use in a power-and-lighting system,
Edison laughed. Edison was using direct current in his system.
He told Tesla very bluntly he was not interested in alternating
current; there was no future to it and anyone who dabbled in
that field was wasting his time; and besides, it was a deadly current whereas direct current was safe. Tesla did not yield any
ground in the discussion—nor could he make any progress in
his effort to get Edison to listen to a presentation of his polyphase
power system. On technical grounds, they were worlds apart.

Nevertheless, because of Batchellor's statement on the valua-
ble work he had done on the Edison direct-current machines in
Europe, Tesla was, without much formality, given a job on
Edison's staff—doing minor routine work. A few weeks later he

had an opportunity to demonstrate his ability. Edison had installed one of his electric-light plants on the steamship *Oregon*, the fastest and most up-to-date passenger ship of that time. The installation worked well for many months but finally both dynamos went out of commission. It was impossible to remove the dynamos and install new ones, so it was necessary to repair the old ones in some way—but this, Edison had been told, was impossible without taking them to the shop. The scheduled sailing date of the ship had passed and Edison was being placed in an embarrassing position over the accumulating days of delay caused by his machines.

Edison asked Tesla if he would go to the ship and see what could be done about the situation. This was in the afternoon. Taking such instruments as he thought he would need, Tesla went aboard the *Oregon*. He found that short circuits had caused some of the armature coils to be burned out; and open circuits had developed elsewhere on the machines.

Calling on members of the crew to assist him, Tesla worked through the night and by 4 A.M. had both machines running as well as they did the day they were newly installed. Walking back to the shop on lower Fifth Avenue at 5 A.M., in the dim early dawn he met a group of men just leaving. In it were Edison, Batchellor, who had returned from Paris in the meantime, and several others who had finished their night's work and were returning to their homes.

"Here is our Parisian running around nights," said Edison.

"Am just coming back from the *Oregon*," Tesla replied. "Both machines are operating."

Edison, amazed, shook his head and turned away without another word. On rejoining the group he said to Batchellor, loud enough for the keen-eared Tesla to hear him, "Batchellor, this is a damn good man."

Thereafter Tesla's status on the staff was raised several levels and he was given closer contact with design and operating problems. He found the work interesting and applied himself to it more than eighteen hours a day, from 10:30 A.M. until 5 A.M., every day including Sundays. Edison, observing his industry, told him, "I have had many hard-working assistants but you take the cake." Tesla observed many ways in which the dynamos could be improved in design to operate more efficiently. He outlined his plan to Edison, and stressed the increased output and lower cost of operating that would result from the changes he suggested. Edison, quick to appreciate the value of increased efficiency, replied, "There's fifty thousand dollars in it for you if you can do it."

Tesla designed twenty-four types of dynamos, eliminating the long-core field magnets then in use and substituting the more efficient short cores, and provided some automatic controls, on which patents were taken out. Months later, when the task was finished, and some of the new machines built and tested and found measuring up to his promises, Tesla asked to be paid the $50,000. Edison replied, "Tesla, you don't understand our American humor." Tesla was shocked to discover that what he thought was a specific promise was being tossed aside merely as a standard practical joke of the day. He received not a penny of compensation from the new designs and inventions, or for the tremendous amount of overtime, beyond the none too generous weekly pay. He resigned his job immediately. This was in the spring of 1885.

In the period of less than a year which he spent with Edison, Tesla had developed a good reputation in electrical circles; so when he was free he was offered an opportunity to capitalize on it. A group of promoters offered to form a company under his name. This looked like a possible chance to bring out his alternating-current system, and he eagerly entered into the project. But when he urged his plan, the promoters informed him they were not interested in alternating current. What they wanted him to develop was a practical arc light for street and factory illumination. In about a year he developed the desired lamp, took out several patents on his invention, and its manufacture and use were under way.

From a technical point of view the venture was a success, but Tesla himself suffered another painful financial experience in connection with it. He had been paid a comparatively small salary during the period of development. According to the agreement, he was to receive his principal compensation in the form of shares of stock in the company. He received a beautifully engraved stock certificate, and then, by some manipulations he did not understand, he was forced out of the company and aspersions were cast upon his ability as an engineer and an inventor. When he sought to convert the certificate into cash, he found that the shares of newly organized companies of undemonstrated power to earn dividends possess very slight value. His opinion of financial men in both the Old World and the New was taking on a decidedly uncomplimentary bias.

Now came the most unpleasant experience of Tesla's life. He was without a source of income, and from the spring of 1886 to the spring of 1887 he was forced to work as a day laborer. "I lived," he said, "through a year of terrible heartaches and bitter tears, my suffering being intensified by material want." Business conditions were none too good in the country. Not only did he have difficulty in getting anyone to listen to his alternating-current project, but even in his effort to earn room and board as a laborer he had tremendous competition, and found it none too easy to secure the most menial tasks at almost starvation wages. He would never discuss this period of his life, probably because it was so unpleasant that he banished all thoughts of it from his memory. Some electrial repair work and even ditch digging at $2 a day were among the jobs he tackled. He resented the utter waste of his abilities more than the personal degradation involved. His education, he said, seemed a mockery.

During the winter of early 1887, while engaged in ditch digging, he attracted the attention of the foreman of the gang who, too, was being forced by circumstances to work below his accustomed level. The foreman was impressed by Tesla's story of his inventions and his great hopes for his alternating-current system. Through this foreman, Tesla said, he was introduced to Mr. A. K. Brown of the Western Union Telegraph Company who put up some of his own money and interested a friend in joining him in Tesla's project.

These two gentlemen organized and financed the Tesla Electric Company, and in April, 1887, established a laboratory at 33–35 South Fifth Avenue (now West Broadway), near Bleecker Street, not far from the shop of the Edison Company. Edison had turned down Tesla's alternating-current idea—and now Tesla was his neighbor with a laboratory of his own, starting to develop the competing idea. Within this small area was to be fought the great battle of the electrical industry over the question of whether direct or alternating current should be used. Edison, already famous, was wholeheartedly committed to direct current; his powerhouses were operating in several cities and, in addition, he had the support of the famous financier, J. P. Morgan. Tesla, on the other hand, was unknown and had only very modest financial support. The direct current was technically simple, whereas alternating current was technically complex. Tesla knew, however, that in these complexities were unlimited possibilities for usefulness.

Tesla's dark days were over. Yet he was soon to discover that the acceptance or rejection of the alternating-current system was not based on technical facts but upon financial considerations, emotional reactions and prejudices, and that human nature was a bigger factor than scientific truths. Nevertheless, in a short time, he would see some of his greatest hopes and dreams realized, and success in large measure reward his efforts.

Once he had achieved something resembling fair conditions under which to carry on his work, the rising star of Tesla's genius shot across the electrical heavens like a meteor. As soon as the newly organized Tesla Electric Company opened its South Fifth Avenue laboratories he started the construction of a variety of pieces of dynamo electric machinery. It was not necessary for him to do any calculating, or work out blueprints. Everything was crystal clear in his mind down to the finest detail of each piece of apparatus. As a result he very quickly produced the working units with which he demonstrated the principles of his polyphase alternating-current system. The single piece of apparatus he had built while in Strassburg, the first model of the induction motor, supplied the physical proof he needed that all the remainder of his calculations were correct.

The apparatuses built in his new laboratory were identical with those which he conceived during the two months in Budapest following the remarkable revelation of the principle of the revolving magnetic field. He did not make the slightest change, he said, in the machines he had mentally constructed at that time. When the machines were physically constructed not one of them failed to operate as he had anticipated. Five years had elapsed since he evolved the designs. In the meantime he had not committed a line to paper—yet he had remembered perfectly every last detail.

Tesla produced as rapidly as the machines could be constructed three complete systems of alternating-current machinery—for single-phase, two-phase and three-phase currents—and made experiments with four- and six-phase currents. In each of the three principal systems he produced the dynamos for generating the currents, the motors for producing power from them and transformers for raising and reducing the voltages, as well as a variety of devices for automatically controlling the machinery. He not only produced the three systems but provided methods by which they could be interconnected, and modifications providing a variety of means of using each of the systems. A few months after opening the laboratory he submitted his two-phase motor to Prof. W. A. Anthony, of Cornell University, for testing. Prof. Anthony reported that it had an efficiency equal to that of the best direct-current motors.

Tesla now not only constructed the machines which he visualized but he worked out the basic mathematical theory underly-

67

ing all of the apparatus. The mathematical theory was so basic that it covered not only the principles applying to machinery for operation at 60 cycles per second, which is the frequency now in standard use, but applied equally well to the whole range of low- and high-frequency currents. With Edison direct current, it had not been found practicable to work with potentials higher than 220 volts on distribution systems; but with alternating current it was possible to produce and transmit currents of many thousands of volts, thus permitting economical distribution, and these could be reduced to the lower voltages for customer use.

Tesla sought to obtain a single patent covering the entire system and all of its constituent dynamos, transformers, distribution systems and motors. His patent attorneys, Duncan, Curtis & Page, filed the application for this patent October 12, 1887, six months after the laboratory opened and five and a half years after Tesla had made his rotary magnetic-field invention.

The Patent Office, however, objected to considering such an "omnibus" application and insisted it be broken down to seven separate inventions, with individual applications filed on each. Two groups of separate applications were filed, on November 30 and December 23 respectively. These inventions were so original and covered such a virgin field of electrical science that they encountered practically no difficulties in the Patent Office and within about six months the patents were issued.*

As a succession of fundamental patents started to issue from the Patent Office to Tesla, the attention of the electrical engineering profession was drawn to this practically unknown inventor. The significance of his epoch-making discoveries was

---

* They were numbered 381,968; 381,969; 381,970; 382,279; 382,280; 382,281 and 382,282. These covered his single and polyphase motors, his distribution system and polyphase transformers. In April of the following year, 1888, he applied for and was later granted five more patents, which included the four- and three-wire three-phase systems. These were numbered 390,413; 390,414; 390,415; 390,721; and 390,820. Within the year he applied for and was granted eighteen more: 401,520; 405,858; 405,859; 416,191; 416,192; 416,193; 416,194; 416,195; 418,248; 424,036; 433,700; 433,701; 433,702; 433,703; 445,207; 455,067; 459,772 and 464,666.

quickly grasped and he was invited to deliver a lecture before the American Institute of Electrical Engineers on May 16, 1888. This invitation was evidence that he had "arrived." Tesla accepted the invitation and put his whole heart into preparing the lecture which, he felt, would enable him to tell the electrical world the magnificent story of his complete alternating-current system and the tremendous advantages it possessed over direct current.

This lecture became a classic of the electrical engineering field. In it Tesla presented the theory and practical application of alternating current to power engineering. This, with his patents, described the foundation, in the matter of circuits, machines and operation, and theory, upon which almost the entire electrical system of the country was established and is still operating today. No new development of anything even slightly approaching comparable magnitude has been made in the field of electrical engineering down to the present time.

Tesla's lecture, and the inventions and discoveries which he included in it, established him before the electrical engineering profession as the father of the whole field of alternating-current power system, and the outstanding inventor in the electrical field.

IT IS not easy to visualize the tremendous burst of electrical development and progress that came out of Tesla's laboratory in the few months after he established it. He produced a tidal wave of advancement which carried the electrical world into the opening of the new power age in one grand surge—although it took several years, naturally, for the commercial exploitation to get under way. The world of electrical engineering was amazed, bewildered and mystified by the host of discoveries thrown into its midst in rapid succession from the Tesla laboratory, and was filled with admiration for the prodigious new genius who had flared up within its ranks.

Tesla's power system, employing high voltage for transmission,

*69*

released electrical powerhouses using direct current from functioning as purely local enterprises, capable of serving an area within a radius of one mile at the very most. His motors used alternating current that could be economically transmitted hundreds of miles, and he provided an economical two- and three-phase system for transmission lines.

The stupendous changes which the Tesla alternating-current inventions and discoveries brought about in the electrical industry can be realized by considering the handicap under which the direct-current powerhouses of the Edison system had operated up to that time. Electricity was generated in powerhouses by relatively small-size dynamos, and the current then distributed to customers over copper conductors laid in conduits under the streets. Some of the electrical energy fed into these conductors at the powerhouse did not arrive as electricity at the far end of the line but was converted along the route to useless heat by the resistance of the conductors.

Electrical energy is composed of two factors, the current, or amount of electricity, and the voltage, or the pressure under which the current is moved. Resistance losses were undergone by the current regardless of the voltage. One ampere of current experienced a definite loss caused by resistance and this loss was the same whether the pressure was 100, or 1,000 or 100,000 volts. If the current value remained fixed, then the amount of energy transported over a wire varied with the voltage. There is, for example, 100,000 times as much energy transported over a wire carrying a current of one ampere at 100,000 volts as there is when the current is one ampere and the pressure is one volt.

If the amount of current carried by a wire is doubled, the heat losses are increased four fold; if the current is tripled, these losses are increased nine fold, and if the current is increased four fold, the losses rise sixteen fold. This situation put definite limits to the amount of current which could be loaded on to conductors.

In addition there is an accompanying drop in pressure. In a

half-mile-long conductor, of the size adopted and under the average currents carried, there would be a drop of about 30 volts. To compensate for this, to some extent, the dynamos were designed to generate 120 volts instead of the standard 110 volts for which lamps were designed. Near the powerhouse the customers would get excess voltage—and a half-mile away their current would be delivered at 90 volts. The early Edison carbon lamps were none too brilliant at 110 volts and gave much less than satisfactory illumination at 90 volts.

As a result of this situation the generation and distribution of direct electric current became very much of a localized matter. The Edison powerhouse could serve an area less than a mile in diameter. In order to give service to a large city it would be necessary to have a powerhouse in every square mile, or even closer if a uniformly satisfactory current were to be supplied. Outside large cities the situation became even more difficult. This was a severe handicap if electricity was to become the universal power source.

Tesla's alternating-current power system, which Edison so emphatically rejected when it was offered to him, freed electricity from its bondage to local isolation. Not alone were his alternating-current motors more simple and flexible than the direct-current machines, but it was possible by a highly efficient method of using transformers, which consisted of two coils of wire around an iron core, to step up the voltage and simultaneously step down the current in a proportionate amount, or use the process in reverse. The amount of energy involved, however, would remain practically unchanged.

Copper wire entails a heavy investment when it is bought by the mile. The diameter of the wire sets the limit to the amount of current it will carry. With the Edison direct-current system there was no practical way for transforming an electric current. The voltage remained fixed and when the current was increased to the carrying capacity of the wire no further expansion was possible on that circuit.

71

With the Tesla system the amount of energy a wire would transport would be increased tremendously by increasing the voltage and letting the current remain fixed below the carrying limit of the circuit. A very small wire could carry a thousand or more times as much electrical energy in the Tesla polyphase alternating system as it could in the Edison direct-current system.

By using Tesla's alternating-current system electricity could be delivered economically at vast distances from the power-house. It would be possible, if desired, to burn coal at the mouth of a mine for generating electricity, and deliver the current cheaply at distant cities, or to generate electricity where water power was available and transmit it to distant points where it could be used.

Tesla rescued the electrical giant from the apron strings of the powerhouse and gave it geographical freedom, the opportunity to expand into the wide-open spaces and work its magic. He laid the foundation for our present superpower system. A development of such magnitude was bound to be loaded with dynamite, and action was sure to follow as soon as someone set a match to the fuse.

## FIVE

TESLA's spectacular lecture and demonstration before the American Institute of Electrical Engineers in New York focused on his work the attention of the electrical fraternity throughout the world. There was no doubt in the mind of the vast majority of electrical engineers that Tesla's discoveries created a new epoch in the electrical industry. But what could be done about it? There were few manufacturers who could take advantage of it. His discoveries were in the same predicament as a ten-pound diamond. No one would question the value of the stone but who would be in a position to purchase it or make any use of it?

Tesla had given no specific thought to commercializing his work at this time. He was in the midst of a program of experimental work which was far from complete and he desired to finish it before engaging in another line of activity. He expected that there would be no alternative to establishing his own company and engaging in the manufacture of his dynamos, motors and transformers. Such a course would take him away from the original experimental work which greatly fascinated him, and which he did not wish to interrupt. Commercializing his inventions, therefore, was a problem that could be postponed, as far as he was concerned, at least as long as the present financing of his work continued.

George Westinghouse, head of the Westinghouse Electric Company in Pittsburgh, was a man of vision. He was already famous as an inventor of numerous electrical devices but principally for his air brake for trains, and had made a fortune out of the exploitation of his own inventions. He recognized the tremendous commercial possibilities presented by Tesla's discoveries and the vast superiority of the alternating- over the direct-current system. He was a practical man of business and was not limited in his choice between the two systems.

Edison, head of the Edison General Electric Company, on the other hand, was under a limitation. Edison's invention was the incandescent electric lamp. Having developed this project, he was faced with finding some way to use it commercially. In order to sell his lamps to the public it was necessary to make the electricity available for lighting them. This necessitated the building of powerhouses and distribution systems. Another kind of electric lamp was already available—the arc lamp—in which he was but slightly interested. The Edison system powerhouses were standardized on low-voltage direct current. At that time direct-current motors were in use, and most technical men believed it was not at all likely there would ever be a practical alternating-current motor. The direct-current system, therefore, offered a number of advantages of a practical nature from Edison's viewpoint.

Westinghouse had no pet project comparable to the incandescent lamp around which he had to throw protecting conditions such as direct-current limitations, so he could look at the Tesla alternating-current discoveries from an unbiased and purely objective point of view. He reached his decision a month after Tesla's lecture. Having done this, he forwarded a brief note to Tesla, making an engagement to see him in the latter's laboratory.

The two inventors had not previously met but each of them was well acquainted with the other's work. Westinghouse, born in 1846, was ten years older than Tesla. He was a short, stout, bearded, impressive-looking individual, and had a habit of directness in conducting his affairs that amounted almost to bluntness. Tesla, thirty-two years old, was tall, dark, handsome, slender and suave. They made a strongly contrasting pair as they stood in Tesla's laboratory, but they had three things in common: they both were inventors, engineers and loved electricity. Tesla had in his laboratory dynamos, transformers, and motors with which he could demonstrate his discoveries and models in actual operating conditions. Here Westinghouse was right at home and quickly became completely sold on the inventor and his inventions.

So favorably impressed was Westinghouse that he decided to act quickly. The story was related to the author by Tesla.

"I will give you one million dollars cash for your alternating-current patents, plus royalty," Westinghouse blurted at the startled Tesla. This tall, suave gentleman, however, gave no outward sign that he had almost been bowled over by surprise.

"If you will make the royalty one dollar per horsepower, I will accept the offer," Tesla replied.

"A million cash, a dollar a horsepower royalty," Westinghouse repeated.

"That is acceptable," said Tesla.

"Sold," said Westinghouse. "You will receive a check and a contract in a few days."

Here was a case of two great men, each possessed with the power of seeing visions of the future on a gigantic panorama, and each with complete faith in the other, arranging a tremendous transaction with utter disregard of details.

The amount involved was unquestionably a record one, for that time, for an invention. While Tesla liked to think of his complete polyphase system as a single invention, he was, nevertheless, selling about twenty inventions on which patents were already issued, and about as many more still to issue. With a total of forty patents involved in the transaction, most of them strongly basic in nature, he received, therefore, about $25,000 per patent. Westinghouse thereby obtained a record-breaking bargain by buying the patents in wholesale quantities.

Westinghouse arranged with Tesla to come to Pittsburgh "at a high salary" for a year, to act as consultant in the commercial application of his inventions. The generous offer made by the Pittsburgh magnate for the purchase of his patents made it unnecessary for Tesla to have any more worries about having to devote a major portion of his time to exploiting his inventions commercially through his own company. He could afford, therefore, to give this year of his time.

The apparatus which Tesla demonstrated to Westinghouse when the latter visited his laboratory, and which worked so beautifully, was designed for operation with a current of 60 cycles. Tesla's investigation had demonstrated that this was the frequency at which the greatest efficiency of operation could be achieved. At higher frequencies there was a saving in the amount of iron required; but the drop in efficiency, and design difficulties that developed, were not compensated for by the very small saving in cost of metal. At lower frequencies the amount of iron required increased, and the apparatus grew in size faster than increased efficiency justified.

75

Tesla went to Pittsburgh and expected to clear up all problems in less than a year. Here, though, he encountered engineers who faced the problem of producing a motor with a design that would insure, first, certainty of smooth and reliable operation; second, economy of operation; third, economy in use of materials; fourth, ease of manufacture; as well as other problems. Tesla had these problems in mind but not with the urgency with which the engineers faced them. In addition he was quite adamant in the choice of 60 cycles as the standard frequency for alternating current while engineers, who had experience on 133 cycles, were not so sure that the lower frequency would be best for the Tesla motors. At any rate there was conflict between the inventor, interested mainly in principles, and engineers interested in practical design problems. Very definite problems were encountered in making the Tesla motor work on a single-phase current in small sizes. In this type of design, artifices had to be incorporated in the motor to achieve some of the characteristics of a two-phase current from the single-phase current that was supplied to operate it.

Tesla was thoroughly disgusted with the situation. He felt his advice concerning his own invention was not being accepted, so he quit Pittsburgh. Westinghouse was sure the situation would work itself out. Seeking to persuade Tesla to remain, he offered him, Tesla revealed many years later, twenty-four thousand dollars a year, one third of the net income of the company and his own laboratory, if he would stay on and direct the development of his system. Tesla, now wealthy and anxious to return to original research, rejected the offer.

Development work proceeded after Tesla left, and soon practical designs were produced for all sizes of motors and dynamos, and their manufacture started. Tesla was happy to note that the 60-cycle standard, his emphatic choice, but which had been questioned on the ground it was less practical in small units, had been adopted as the standard frequency.

On returning to his New York laboratory, Tesla declared that he had not made a single worth-while contribution to electrical science during the year he spent at Pittsburgh. "I was not free at Pittsburgh," he explained; "I was dependent and could not work. To do creative work I must be completely free. When I became free of that situation ideas and inventions rushed through my brain like a Niagara." During the following four years he devoted a large fraction of his time to further developments of his polyphase power system, and applied for, and was granted, forty-five patents. Those granted in foreign countries would bring the total to several times this number.

THE ideas of the two giants among inventors—Edison and Tesla—were meeting in head-on battle. Out of the laboratories of the two geniuses, within sight of each other in South Fifth Avenue in New York, had come world-shaking developments.

There had been considerable conflict between Edison, who adhered strictly to direct current, and those who supported the claims for alternating current. The Thomson-Houston Company and the Westinghouse Electric Company had extensively developed this field for series electric lighting and arc lighting before the Tesla power system was developed. Edison had engaged in many tilts at these competitors, attacking alternating current as unsafe because of the high voltages used. The advent of the Tesla system added fuel to the fire.

It was Tesla's belief that when the New York State Prison authorities adopted high-voltage alternating current for electrocution of condemned prisoners, the Edison interests had engineered the project to discredit alternating current. There is no doubt about the aid the prison authorities' choice gave the direct-current group; but their decision was undoubtedly based on the fact that direct current could not, by any practical means, be produced at the high voltages required, whereas alternating-current potentials could be very easily increased. Direct current is just as deadly, at the same voltage and amperage, as alternating current. In this "war of the currents," however, as in other wars, appeals to the emotions, instead of to simple facts, were the governing influences.

The task of putting the United States on an electrical power basis—which is what George Westinghouse undertook when he began to exploit the Tesla patents—was a gigantic one requiring not only engineering talent but capital. The Westinghouse Electric Company experienced a tremendous expansion in the volume of its business, but the upward surge came at a time when the country was going into a stage of commercial and financial depression; and Westinghouse soon found himself in difficulties.

This was, in addition, an era in which competing giant financial interests were battling for control of the industrial structure of the country through control of capital. It was a time of mergers, a period when the financial interests were building larger units of production by uniting smaller companies in related fields, frequently forcing these combinations without regard to what the owners of the companies desired.

One merger, internally initiated and arranged by mutual consent, brought together the Thomson-Houston Company and the Edison General Electric Company, the two biggest competitors of Westinghouse Electric, to form the present General Electric Company. This was a challenge to competing financial interests.

Westinghouse had expanded his business at a very rapid rate in exploiting the Tesla patents. Because his financial structure thereby lost a certain amount of flexibility, he became vulnerable to financial operators and soon found himself in the toils of a merger that involved uniting several other small companies with his organization. Financial interests that had stepped into the situation demanded that the Westinghouse Electric Company be reorganized as a step toward bringing about a merger with it of the U. S. Electric Company and the Consolidated Electric Light Company, the new unit to be known as the Westinghouse Electric and Manufacturing Company.

Before this reorganization would be consummated the financial advisers, in strategic positions, insisted that Westinghouse jettison some of his plans and projects which they considered inadvisable or a detriment to getting the new company onto a new foundation that would be sounder from a financial point of view.

One of the requirements was that Westinghouse get rid of the contract with Tesla calling for royalty payments of $1 per horsepower on all alternating-current articles sold under his patents.* The financial advisers pointed out that if the business which Westinghouse expected the company would do under the Tesla patents in the ensuing year was anywhere near as great as estimated, the amount to be paid out under this contract would be tremendous, totaling millions of dollars; and this, at the time of reorganization, appeared a dangerous burden, imperiling the stability which they were trying to attain for the new organization.

Westinghouse strenuously objected to the procedure. This patent-royalty payment, he insisted, was in accordance with usual procedures and would not be a burden on the company, as it was included in costs of production, was paid for by the customers, and did not come out of the company's earnings. Westinghouse, himself an inventor of first magnitude, had a strong sense of justice in his dealings with inventors.

The financial advisers, however, were not to be overruled. They nailed Westinghouse on the spot by insisting that the million dollars he had paid Tesla was more than adequate compensation for an invention, and that by making such an exorbitant payment he had imperiled the financial structure of his company and jeopardized his bankers' interest. Any further imperiling of the reorganization by any effort to retain the royalty contract would, it was argued, result in the withdrawing of support that would save the company.

* No documentary evidence exists concerning this contract. The author located two sources of information. One was in complete agreement with the story here related. The other states that the million-dollar payment was advance royalties and Tesla so described it to him, declaring no further royalties were paid.

79

The situation boiled down to the common "Either-Or" technique.

Westinghouse was required to handle the negotiations with Tesla. No situation could be more embarrassing to him. Nevertheless, Westinghouse was a realist among realists. He never hesitated to face facts squarely and with a blunt directness. "I will give you one million dollars cash for your alternating-current patents, plus royalty": he had been both brief and blunt when he purchased the patents from Tesla. Now he was faced with the problem of undoing the situation into which he had entered with such brevity. Then money talked and he held the money. Now Tesla held the dominant position; he held a perfectly valid contract worth many millions, and he could go to court to force compliance with its terms. Edison's successful suit against infringers of his electric-light patent, bringing disaster to many companies that violated his patent property rights, had caused the whole industrial world to hold a new and wholesome respect for patent rights.

Westinghouse had no reason for believing that Tesla would show the slightest inclination to relinquish his contract or permit its terms to be changed to provide a smaller rate of royalty. He knew that Tesla's pride had been hurt by the disagreement with the Pittsburgh engineers, and that he might not now be in a conciliatory mood. On the other hand, Westinghouse knew that he had succeeded in having Tesla's ideas adopted. His greatest comfort came from the fact that he had entered into the contract with good faith—and with the same good faith he was trying to handle a much less satisfactory situation. Perhaps he could offer Tesla an executive position in the company in lieu of the contract. There would be mutual advantages in such an arrangement.

There is no means of fixing the definite value of the contract Tesla held. His patents covered every department of the new alternating-current power system, and royalties could be collected on powerhouse equipment and motors. At that time the electric power industry had barely started; no one could look into the future and see the tremendous volume of business that would be developed.*

It would be a tough job for any executive, no matter how shrewd or clever, to talk a man out of a contract that would net him many millions of dollars, or induce him to accept a reduction in rates amounting to millions.

Westinghouse called on Tesla, meeting him in the same South Fifth Avenue laboratory where he had purchased the patents four years before. Without preliminaries or apologies Westinghouse explained the situation.

"Your decision," said the Pittsburgh magnate, "determines the fate of the Westinghouse Company."

"Suppose I should refuse to give up my contract; what would you do then?" asked Tesla.

"In that event you would have to deal with the bankers, for I would no longer have any power in the situation," Westinghouse replied.

"And if I give up the contract you will save your company and retain control so you can proceed with your plans to give my polyphase system to the world?" Tesla continued.

* The latest data available indicate that in 1941 there was 162,000,000 horse-power of electrical generating machinery in operation in the United States, practically all of it for alternating current. Assuming a uniform growth from 1891 to 1941, the installed horsepower in 1905, when the first Tesla patents would have expired, would have been about twenty million. This figure is, apparently, too high.

According to a census of central stations in the United States conducted by T. Commerford Martin (*Electrical World*, March 14, 1914) the horsepower of generators in operation in 1902 was 1,620,000 and in 1907 the figure had risen to 6,900,000. On a pro rata, per-year basis, this would make the figure for 1905, the year when Tesla's first patents expired, 5,000,000. During this period many manufacturers who had been using steam power installed dynamos in their factories and operated isolated plants. These would not be included in the central-station figures and, if added, would bring the total horsepower to perhaps 7,000,000. Tesla would have been entitled to $7,000,000 royalties on this equipment, on the basis of his $1-per-horsepower arrangement. In addition he would have been entitled to royalties on motors that used the power generated by these dynamos. If only three quarters of the current generated were used for power, this would have entitled him to additional royalties of $5,000,000, or a total of $12,000,000.

"I believe your polyphase system is the greatest discovery in the field of electricity," Westinghouse explained. "It was my efforts to give it to the world that brought on the present difficulty, but I intend to continue, no matter what happens, to proceed with my original plans to put the country on an alternating-current basis."

"Mr. Westinghouse," said Tesla, drawing himself up to his full height of six feet two inches and beaming down on the Pittsburgh magnate who was himself a big man, "you have been my friend, you believed in me when others had no faith; you were brave enough to go ahead and pay me a million dollars when others lacked courage; you supported me when even your own engineers lacked vision to see the big things ahead that you and I saw; you have stood by me as a friend. The benefits that will come to civilization from my polyphase system mean more to me than the money involved. Mr. Westinghouse, you will save your company so that you can develop my inventions. Here is your contract and here is my contract—I will tear both of them to pieces and you will no longer have any troubles from my royalties. Is that sufficient?"

Matching his actions to his words Tesla tore up the contract and threw it in the waste basket; and Westinghouse, thanks to Tesla's magnificent gesture, was able to return to Pittsburgh and use the facilities of the reorganized company, which became the present Westinghouse Electric and Manufacturing Company, to make good his promise to Tesla to make his alternating-current system available to the world.

Probably nowhere in history is there recorded so magnificent a sacrifice to friendship as that involved in Tesla's stupendous gift to Westinghouse of $12,000,000 in unpaid royalties, although Westinghouse personally received only indirect benefits from it.

It is also probable that the failure to pay Tesla these royalties resulted in one of the greatest handicaps to scientific and industrial progress which the human race has experienced. A few years later Tesla, still an intellectual giant far from the peak of his greatest growth, still pouring forth a profusion of inventions and discoveries of first magnitude, equal in importance to his first efforts which put the world on an electrical power basis, found himself without funds with which to develop his discoveries, with the result that many of them have been lost.

Nearly fifty years after this majestic relinquishment of wealth on the altar of friendship, during which time Tesla had had opportunity to see the United States and the world as a whole wax wealthy out of the power he had made available, he was called on to respond, with a speech, to honorary citation by the Institute of Immigrant Welfare. Tesla, then about eighty, was unable to appear in person. He had experienced decades of poverty in which he faced ridicule for his failure to develop inventions which he declared he had made, and had been forced to move frequently from hotel to hotel, owing to inability to pay his bills. In spite of these experiences he developed no rancor toward Westinghouse in whose behalf he sacrificed his $12,000,000 in royalties. Instead, he retained his original warm friendship. This is indicated by a statement in the speech he sent to the Institute to be read at its dinner held in the Hotel Biltmore, May 12, 1938:

"George Westinghouse was, in my opinion, the only man on this globe who could take my alternating-current system under the circumstances then existing and win the battle against prejudice and money power. He was a pioneer of imposing stature, one of the world's true noblemen of whom America may well be proud and to whom humanity owes an immense debt of gratitude."

# SIX

---

WHEN Tesla left the Westinghouse plant at Pittsburgh in 1889 to return to his laboratory in New York, he entered a new world. The magnificent polyphase system which he had already produced was but a small sample of the greater wonders that still remained to be revealed, and he was anxious to start exploring the new realm.

He was not approaching an entirely unknown realm in which he would have to feel his way in darkness in the hope of stumbling upon something of value, although anyone else at that time would have been in that position. On that fateful afternoon in February in Budapest in 1882, when he was given the vision of the rotating magnetic field, there had come with it an illumination that revealed to him the whole cosmos, in its infinite variations and its myriad of forms of manifestations, as a symphony of alternating currents. For him, the harmonies of the universe were played on a scale of electrical vibrations of a vast range in octaves. In one of the lower octaves was a single note, the 60-cycle-per-second alternating current, and in one of the higher octaves was visible light with its frequency of billions of cycles per second.

Tesla had in mind a course of experimentation in which he would explore this region of electrical vibration between his alternating current and light waves. He would increase the frequency of the alternating current through the unknown intervening regions. If one note in a lower octave produced such a magnificent invention as the rotating magnetic field and the polyphase system, who could imagine the glorious possibilities that lay hidden on other notes in higher octaves? And there were thousands of octaves to be explored. He would construct an electrical harmonium by producing electrical vibrations in all frequencies, and study their characteristics. He would then, he hoped, be able to understand the motif of the cosmic symphony of electrical vibrations that pervaded the entire universe.

Tesla, at the age of thirty-three, was now wealthy. He had received $1,000,000 from the Westinghouse Company for his first crop of inventions. Of this, $500,000 went to A. K. Brown and his associate who had financed his experiments. Still greater inventions were to follow. He would never need money. He would,

he then believed, have royalties in the millions from his alternating-current patents. He could spend as freely as he wished,
penetrating the secrets of Nature and applying his discoveries to
human welfare. It was his responsibility to be so engaged. He
knew he was gifted as no other man had been blessed with vision,
talent and ability; and he in turn would endow the world with
supernal treasures of scientific knowledge which he would extract from the secret recesses of the universe and, through the
activities of his mighty mind, transform into agencies to brighten
the lives, lighten the labors and increase the happiness of the
human race.

Was he a superegoist in his attitude? If so, he was not activated
by selfish motives. To him it mattered not what he thought, so
long as he remained objective in his thinking and his thoughts
could be translated into demonstrable facts. What if he did consider himself greater than other men: did not this viewpoint
conform to the facts? Suppose he did consider himself a man of
destiny. Could he not bring evidence to support the contention?
It was not necessary for Tesla actually to see an event occur in
order to enjoy its realization. Had he not as a youth declared that
he would make a practical alternating-current motor, only to be
told by his professor that the goal was impossible of attainment
—and had he not already accomplished this "impossibility"?
Had he not taken the direct-current dynamos of Edison, whom
all the world looked upon as a great genius, and had he not greatly
greatly improved their design and operation; and in addition,
had he not produced a vastly superior system for producing, distributing and using electricity? To all of these inquiries Tesla
could answer in the affirmative without going beyond the bounds
of modesty concerning his achievements.

His attitude was not that of an egoist. It was an attitude of
supreme faith in himself and in the vision that had been given
him. To a man of ability, with such supreme faith in himself,
and necessary financial resources to advance his purposes, the
world of accomplishments is without limits. This was the picture

85

of Tesla as he returned to his laboratory in lower Fifth Avenue, New York, in the latter part of 1889.

Tesla had studied a wide range of frequencies of alternating current in order to select the frequency at which his polyphase system would operate most efficiently. His calculations indicated important changes in characteristics and effects as the frequency of the current was increased; and his observations with the electrical machinery he built confirmed his calculations. He noted that ever smaller quantities of iron were required as the frequencies were increased, and he now wished to explore the very high frequencies at which unusual effects should be produced without any iron in the magnetic circuit.

When, back in Budapest following his rotating magnetic-field discovery, he had played with mental calculations of the properties of alternating currents all the way from the very lowest frequency up to that of light, no one had yet explored this region. James Clerk Maxwell, at Cambridge University, England, had, however, nine years before, in 1873, published his beautiful presentation on an electromagnetic theory of light, and his equations indicated that there was a vast range of electromagnetic vibrations above and below visible light—vibrations of much longer and much shorter wavelengths. While Tesla was engaged in making models of his polyphase system in 1887, too, Professor Heinrich Hertz, in Germany, put the Maxwell theory to test in the range of waves a few meters long. He was able to produce such waves by the spark discharge of an induction coil, and was able to absorb such waves from space and change them back to a small spark at some distance from the coil.

Hertz's work gave support to Tesla's theory that there was an interesting discovery to be made on almost every note of the whole gamut of vibrations between the known ones of the electrical current and those of light. Tesla felt sure that if he could continually increase the frequency of electrical vibrations until they equaled that of light, he would be able to produce light by a direct and highly efficient process instead of the extremely wasteful process used in the Edison incandescent lamp, in which the useful light waves were a very small fraction of the wasted heat waves emitted in the process, and only five per cent of the electrical energy was effectively utilized.

Tesla started his investigations by building rotary alternating-current dynamos with up to 384 magnetic poles, and with these devices he was able to generate currents up to 10,000 cycles per second. He found that these high-frequency currents presented many fascinating possibilities for even more efficient power transmission than his very practical 60-cycle polyphase system. He therefore carried on a parallel line of research into transformers for raising and lowering the voltage of such currents.

High-frequency alternating-current dynamos, similar to those designed by Tesla in 1890, were subsequently developed by F. W. Alexanderson into the high-power wireless transmitters which put transatlantic wireless transmission, more than two decades later, on such a sound practical basis that the Government would not permit control of it to go to a foreign country and preserved for the United States its predominant position in world wireless.

The high-frequency current transformers which Tesla developed proved to be spectacular performers. They contained not a trace of iron; as a matter of fact, the presence of iron was found to interfere with their operation. They were air-core transformers and consisted merely of concentric primary and secondary coils. The voltages he was able to produce with these transformers, which became known as Tesla coils, were very high. In the early experiments he attained potentials that would spark across a couple of inches of air, but in a short time he made tremendous progress and was producing flaming discharges. In working with these voltages he encountered difficulties in insulating his apparatus, and so he developed the technique that is now in universal use in high-tension apparatus: that of immersing the apparatus in oil and excluding all air from the coils, a discovery of great commercial importance.

There was a limit, however, above which the use of rotary

generators of high-frequency currents was not practicable, so Tesla set about the task of developing a different type of generator. There was nothing novel about the basic idea he employed. In rotary dynamos, current is generated by moving a wire in a circle past a number of magnetic poles in succession. The same effect can be attained by moving the wire back and forth with an oscillating motion in front of one magnetic pole. No one, however, had as yet produced a practical reciprocating dynamo. Tesla produced one that was extremely practical for his particular purpose; but otherwise it had little utility, and he later felt that he could have employed much better the time he spent on it. It was an ingenious single-cylinder engine without valves, and could be operated by compressed air or steam. It was supplied with ports like a small two-cycle marine engine. A rod extended from the piston through the cylinder head at either end, and at each end of the rods was attached a flat coil of wire which, by the reciprocating action of the piston, was caused to move back and forth through the field of an electromagnet. The magnetic field through its cushioning effect served as a flywheel.

Tesla was able to obtain a speed of 20,000 oscillations per minute, and to maintain such a remarkable degree of constancy in operation that he proposed the maintenance of equally constant speed of operation for his 60-cycle polyphase system and the use of synchronous motors, geared down to the proper extent, as clocks which would furnish correct time wherever alternating current was available. This proposal furnished the foundation for our modern electric clocks. As with many another of his practical and useful suggestions, he did not take out a patent on the idea, and gained no financial advantage from it.

IN WORKING with his polyphase system, Tesla gained a thorough understanding of the part played by the two factors, capacity and inductance, in alternating-current circuits; the former acting like a spring and the latter like a storage tank. His calculations indicated that with currents of sufficiently high frequency it would be possible to produce resonance with relatively small values of inductance and capacity. Producing resonance is tuning a circuit electrically. The mechanical effects analogous to electrical resonance are the causing of a pendulum to swing through a wide arc by giving it a series of very light but equally timed touches, or the destruction of a bridge by soldiers marching in unison over it. Each small vibration re-enforces its predecessors until tremendous effects are built up.

In a tuned electrical circuit a condenser supplies the capacity and a coil of wire supplies the inductance. A condenser ordinarily consists of two parallel metal plates separated from each other a short distance by an insulating material. Each plate is connected to either end of the inductance coil. The size of the condenser and the coil is determined by the frequency of the current. The coil-condenser combination and the current are tuned to each other. The current can be pictured as flowing into the condenser until it is fully charged. It then flows elastically into the inductance coil, which stores the energy by building up its magnetic field. When the current ceases to flow in the coil, the magnetic field collapses and gives back to the coil the energy previously used in building up the magnetic field, thus causing a current to flow back into the condenser to charge it up to overflowing again, so that it is ready to repeat the process. This flow back and forth between the condenser and coil takes place in step with the periodic reversal of the alternating current which supplies the energy when resonance is established. Each time it takes place, the charging current comes along at the right instant to give it a boost, so that the oscillations build up to tremendous values.

Tesla, in discussing this plan of electrical tuning of circuits in a lecture, given several years later, said:

The first question to answer then is whether pure resonance effects are producible. Theory and experiment show that such is impossible in nature for, as the oscillations become more vigorous, the losses in vibrating bodies and environing media rapidly increase, and neces-

*89*

sarily check the vibrations, which would otherwise go on increasing forever. It is a fortunate circumstance that pure resonance is not producible for, if it were, there is no telling what dangers might lie in wait for the innocent experimenter. But, to a certain degree, resonance is producible, the magnitude of the effects being limited by the imperfect conductivity and imperfect elasticity of the media, or, generally stated, frictional losses. The smaller these losses the more striking are the effects.

Tesla applied the electrical tuning principles to his coils and discovered that he was able to produce tremendous resonance effects and build up very high voltages. The tuning principles he developed in 1890 are those which have made our modern radio, and the development of the earlier art, "wireless," possible. He had been working with, and demonstrating, these principles before others who received credit had begun to learn the first lessons in electricity.

Seeking a new source of high-frequency currents, higher than could be produced by any mechanical apparatus, Tesla made use of a discovery that had been made the year in which he was born, by Lord Kelvin, in England, in 1856, and for which no use had thus far been found. Up to the time of Kelvin's discovery it had been believed that when a condenser was discharged the electricity flowed out of one plate into the other, like water being poured from a glass, thus establishing equilibrium. Kelvin showed that the process was far more interesting and complex; that its action was like the bobbing up and down that takes place when a weighted stretched spring is released. The electricity, he showed, rushes from one plate into the other and then back again, the process continuing until all of the stored up energy is used up in overcoming frictional losses. The back-and-forth surges take place at a tremendously high frequency, hundreds of millions a second.

The combination of condenser discharges and tuned circuits opened a new realm in electrical science as significant and as important as Tesla's polyphase system. He worked out remarkably simple and automatic methods for charging the condensers by low voltage (direct and alternating currents), and discharging them through his new air-core transformers, or Tesla coils, to produce currents of enormously high voltages that oscillated at the tremendously high frequency of the condenser discharge. The properties of these currents were unlike anything that had been seen before. He was again pioneering in an entirely new field, with tremendous possibilities. He labored feverishly in his laboratory; and as he lay in bed at night for his five-hours' rest, which included two hours of sleep, he formulated new experiments.

Tesla announced the heating effect of high-frequency currents on the body in 1890 and proposed their use as a therapeutic device. In this he was a pioneer, but soon had many imitators here and abroad who claimed to be originators. He made no effort to protect his discovery or prevent the pirating of his invention. When the same observation was made thirty-five years later in laboratories using vacuum-tube oscillators as the source of the high-frequency currents, it was hailed as a new discovery and developed as a modern wonder. Tesla's original discovery is, however, the basis of a vast array of very recent electronic applications in which high-frequency currents are used to produce heat for industrial purposes.

When he gave his first lecture on the subject before the American Institute of Electrical Engineers at Columbia College, in May, 1891, he was able to produce spark discharges five inches long, indicating a potential of about 100,000 volts, but, more important, he was able to produce phenomena which included electrical sheets of flame, and a variety of new forms of illumination—electric lamps the like of which had never been seen before, nor dreamed of in the wildest imagination of any experimenter.

This lecture produced a sensation in engineering circles. He was already famous in this field for the astounding revelations he had made before the same organization on that earlier occasion when he described his discovery of the polyphase alternating-current system. That discovery was an intellectual accomplishment of bewildering brilliance, made impressive by the tremendous commercial importance of the discovery. The experiments with the high-frequency and high-potential currents, however, were spectacular; the crackling of the high-voltage sparks, the flashing of the high-potential sheets of electrical flame; the brilliant bulbs and tubes of electrical fire, the amazing physical effects he produced with the new currents, made a profound emotional appeal to the startled beholders.

The man who could produce these two pioneering developments within two years must be more than a genius! The news of his new accomplishment flashed quickly throughout the world, and Tesla's fame now rested on a double foundation.

The world-wide fame that came to him at this time was unfortunate. Tesla would have been entirely superhuman had he not derived a great deal of satisfaction out of the hero-worshiping adulation that now came to him. It was only five years ago that he had been hungry and penniless in the streets of New York, competing with equally hungry hordes of unemployed for the few existent jobs calling for brute labor, while his head bulged with important inventions which he was anxious to give to the world. No one would listen to him then—and now the intellectual élite of the nation were honoring him as an unrivaled genius.

TESLA was a spectacular figure in New York in 1891. A tall, dark, handsome, well-built individual who had a flair for wearing clothes that gave him an air of magnificence, who spoke perfect English but carried an atmosphere of European culture which was worshiped at that time, he was an outstanding personality to all who beheld him. Hidden behind his quiet, self-effacing demeanor, and an extreme modesty that manifested itself as an exaggerated shyness, was the mind of a genius which had worked electrical wonders that fired the imagination of all and exceeded the understanding of the vast majority of the population. In addition Tesla was a young man, not yet thirty-five, who had recently received a million dollars and was a bachelor.

A bachelor with a million dollars, culture and fame, could not avoid being a shining mark in New York in the early years of the gay nineties. Many were the designing matrons with marriageable daughters who cast envious eyes in the direction of this eligible young man. The social leaders looked upon him as a fascinating decoration for their salons. The big men of business looked upon him as a good man to know. The intellectuals of the day found his almost unbelievable accomplishments a source of inspiration.

Except at formal dinners Tesla always dined alone, and never under any circumstances would he dine with a woman at a twosome dinner. No matter how much a woman might gush over him or strive to gain his favor, Tesla, in most adamant fashion, maintained a thoroughly impersonal attitude. At the Waldorf-Astoria and at Delmonico's he had particular tables which were always reserved for him. They occupied secluded positions in the dining rooms because when he entered either room he was the cynosure of all eyes and did not enjoy being on exhibition.

In spite of all of the adulation that was heaped upon him, Tesla had but one desire—to continue his laboratory experiments undisturbed by outside distractions. There was a tremendous empire of new knowledge to be explored. He was fired with a potential of enthusiasm for the work that was as high as the voltage of the currents with which he was working, and new ideas were coming to him with almost the rapidity of the cycles in his high-frequency current.

There were three broad fields in which he wished to develop applications which were now clearly outlined in his mind: a system of wireless power transmission that would excel his own polyphase system, a new type of illumination, and the wireless transmission of intelligence. He wished to work on them all si-

*93*

multaneously. They were not separate and isolated subjects but all closely intermeshed, all notes on that vast cosmic scale of vibration represented by his beloved alternating currents. He did not wish to play on one note at a time, as would a violinist; he preferred to play as a pianist, striking many notes at once and weaving them into beautiful chords. Were it possible to occupy the position of leader and simultaneously play all of the instruments in a great symphony orchestra, he would have been still better pleased. The instruments in his orchestra, however, would be electrical devices oscillating in tune with their energizing currents or with their environment. To the extent that he was unable to realize his most expansive desires, he was under mental pressure that drove him to a working pace which no individual of ordinary strength could withstand without a resulting complete physical breakdown.

The spectacular lecture and demonstration on high-frequency and high-potential currents which he gave before the American Institute of Electrical Engineers in February, 1891, at Columbia College, created as profound a sensation as did his earlier one. Each opened an entirely new realm of scientific investigation and practical discoveries. The discoveries contained in either lecture would have been sufficient to stand as the fruit of a lifetime's work and bring lasting fame. Two such events in rapid succession seemed almost unbelievable—yet Tesla seemed to be scarcely well launched on his career, with more important work still to come.

Requests that he give lectures came from learned societies throughout this country and Europe, but he begged to be excused because of the tremendous pressure on his time which his work entailed. Equally insistent were the social demands that were being made upon him. Social groups sought in every way to honor him, and incidentally to shine in his reflected glory. Tesla was not vulnerable to the importunings of the socialites who sought him merely as a scintillating satellite, but the clever "lion hunters" of that day soon discovered his Achilles' heel—an intelligent interest in his accomplishments and a sympathetic ear for his dreams of wonders still to come.

With this technique in successful operation, Tesla was captured and soon completely lionized. He was guest of honor at a continuous round of functions and he met the social obligations involved in them by staging, in return, elaborate dinners at the Waldorf-Astoria followed by demonstration parties at his laboratory on South Fifth Avenue. Tesla never did a halfway job on anything. When he staged a dinner he left nothing to chance in the matter of cuisine, service and decorations. He sought rare fish and fowl, meats of surpassing excellence, and choicest liquors and exquisite wines of the best vintages. His dinners were the talk of the town and having been a guest at a Tesla dinner was a mark of social distinction, proof of membership in the inner group of the élite within Ward MacAllister's "400." At these dinners Tesla presided as a most meticulous host, or more accurately, as an old-world absolute monarch, for he would sample all food brought to the dining room; and rarely did an event pass without the grandiose host sending back some sauce or wine of unquestioned excellence as unworthy of his guests.

Following each of these meals Tesla would escort his guests to his laboratory below Washington Square; and here his demonstrations were even more spectacular than his dinners. He had a flair for the dramatic; and the strange-looking devices with which his laboratory was furnished provided a grotesque and bizarre background for the fantastic displays of seemingly unearthly forces that with invisible fingers set objects whirling, caused globes and tubes of various shapes to glow resplendently in unfamiliar colors as if a section of a distant sun were suddenly transplanted into the darkened room, and crackling of fire and hissing sheets of flame to issue from monster coils to the accompaniment of sulfurous fumes of ozone produced by the electrical discharges that suggested this magician's chamber was connected directly with the seething vaults of hell. Nor was this illusion dispelled when Tesla would permit hundreds of thousands of volts of electricity to pass through his body and light a lamp or melt a wire which he held.

The amazing feat of harmlessly passing through his body currents of tremendously high voltage and high frequency was one which Tesla evolved by his mental experiments long before he had an opportunity to test them in his laboratory. The low-frequency alternating currents, such as are now used on home-lighting circuits, would, he knew from unpleasant experiences, produce a painful shock if passed through the body. When light waves impinged on the body, however, no such painful sensation was produced. The only difference between the electric currents and light waves, he reasoned, was a matter of frequency, the electric currents oscillating at the rate of 60 per second and the light waves at billions per second.

Somewhere between these two extremes the shock-producing property of electromagnetic vibrations must disappear; and he surmised the point would be near the lower end of the gap. Damage done to the body by electric shock he divided into two factors, one—the destruction of tissues by the heating effect which increased or diminished as the amperage of the current was raised or lowered; and two—the sensation of acute pain which varied with the number of alternations of the current, each alternation producing a single stimulus which was transmitted by the nerves as a pain.

Nerves, he knew, could respond to stimuli up to a rate of about 700 per second, but were unable to transmit impulses received at a more rapid rate. In this respect they acted very much like the ear, which is unable to hear air vibrations above a frequency of about 15,000 per second, and the eye, which is blind to color vibrations of a frequency higher than that in violet light.

When he constructed his high-frequency alternating-current dynamos, he had frequencies up to 20,000 per second with which to test his theory; and by finger tests across the terminals he was able to demonstrate that the nerves were unable to perceive the individual vibrations at this rapid rate. The amperage, which carried the tissue-destroying power, was still too high in the output of these machines to pass safely through his body, even though the sensation of pain was lacking.

By passing these currents through his newly invented air-core transformers, he could increase their voltage ten-thousand fold and reduce the amperage proportionately. The current density would thereby be reduced below the point at which it would injure tissues. He would then have a current which would not produce sensation and would not harm the tissues. He cautiously tested the theory by passing the currents through two fingers, then his arm, next from hand to hand through his body and finally from his head to his feet. If a spark jumped to or from his body, there was a pin-prick sensation at the point of contact, but this could be eliminated by holding a piece of metal to and from which the spark could jump while the current passed through the tissues without producing any sensation.

The energy content of these currents, which is proportionate to the current multiplied by the voltage, could be very high and produce spectacular effects such as melting metal rods, exploding lead disks, and lighting incandescent or vacuum-tube lamps after passing painlessly through his body.

The European scientific societies were persistent in their efforts to induce Tesla to accept their invitations to lecture before them, and finally he acceded. He set extravagantly high standards for the contents of his lectures, and their preparation entailed a tremendous amount of labor. All of the material had to be entirely new. He would never repeat an experiment previously presented. Every technical statement had to be tested at least twenty times to insure complete accuracy. His lectures would last two or three hours; and every minute of the time was crowded with new and awe-inspiring demonstrations of his constant stream of discoveries. He used a great array of devices fashioned by himself and built in his own laboratories to illustrate

*97*

his talks. A Tesla lecture, therefore, was an extremely important event in the scientific world and a most impressive occasion to those who were fortunate enough to be able to attend.

Tesla arranged to give a lecture before the Institution of Electrical Engineers in London on February 3, 1892, and one before the International Society of Engineers in Paris on February nineteenth. His decision to give the European lectures was influenced to some extent by the fact that they would afford him an opportunity to visit his home in Gospic, for recent letters had indicated that his mother's health was failing.

The lecture before the Institution of Electrical Engineers was a great success. English engineering journals, as will be seen, had been niggardly in extending recognition to Tesla for priority in the discovery of the rotating magnetic field, and had belittled the practical value of his polyphase alternating-current system, but in this attitude they were not representative of the great body of engineers, who were most generous in their praise and enthusiasm; and the attitude of the engineers was shared by the English scientists.

When Tesla arrived in London he was entertained at many places by famous men. At the Royal Institution, where the immortal Michael Faraday had carried on his fundamental researches in magnetism and electricity, Sir James Dewar, and a committee of equally famous scientists, sought to prevail upon Tesla to repeat his lecture before that organization. Tesla could be plain stubborn in sticking to his plans, and in this case was exhibiting his usual firmness. The famous Scottish scientist matched Tesla's stubbornness with an equal persuasive persistence. He escorted Tesla to Faraday's chair, an almost sacred relic to English science, seated him in this throne, and then brought out an almost equally precious heirloom, a portion of a bottle of whiskey, the remainder of Faraday's personal supply, untouched for nearly a quarter of a century. Out of this he poured a generous half glass for Tesla. Sir James won. Tesla relented and gave the lecture the following evening.

Lord Rayleigh, the eminent English physicist, was chairman of the meeting at the Royal Institution, which was attended by the élite of the scientific world and a generous representation of the nobility of the realm. Rayleigh, after witnessing the performance of Tesla's experiments, which were none the less awe inspiring to scientists than to laymen, showered words of praise on the inventor.

Rayleigh declared that Tesla possessed a great gift for the discovery of fundamental scientific principles, and urged that he concentrate his efforts on some one big idea.

Tesla, in his conversation after the meeting, disclaimed ability as a great discoverer; but in this he was merely being modest, for he knew that he was unique among men in his ability to discover fundamental truths. He did, however, give very serious consideration to Rayleigh's suggestion that he concentrate on some one big idea. It is doubtful, however, whether Rayleigh's suggestion was good advice. Tesla's mind had a range that was cosmic in magnitude and adjusted to broad slashing advances through unknown regions. Rayleigh's advice was like suggesting to an explorer who had unique ability for penetrating an unknown continent and opening it to civilization that he settle down and cultivate a homestead, since that would give more definite and specific returns for efforts expended.

Two weeks later Tesla gave his scheduled lecture before the Physical Society in Paris and repeated it before the International Society of Electrical Engineers. This was his second visit to Paris since he had quit his job with the Continental Edison Company in that city eight years before. Immediately after leaving the Westinghouse Company in the autumn of 1889—at which time, too, he completed his U. S. citizenship requirements—he had made a brief visit to Paris to attend the International Exposition. In the meantime, the fame of his polyphase system had spread to Europe; and to this was added the glory for his spectacular work with the new high-frequency currents. He was given a hero's reception in Paris, as well as in London.

It would be interesting to know what thoughts passed through the minds of the executives of the Continental Edison Company as they observed the tremendous contributions to science and industry by the engineer whose services they had lost through their penny-wise tactics when they were offered in 1883, and could undoubtedly have purchased for a relatively small amount, the polyphase system for which Westinghouse paid Tesla $1,000,000 five years later.

A TESLA lecture was an avalanche of new and fascinating electrical knowledge. He completely overwhelmed his listeners with a wealth of spectacular original experiments, and as a result almost every individual contribution lost its identity in the dazzling concentration of the whole galaxy of startling developments.

In the 1892 lectures, entitled "Experiments with Alternating Currents of High Potential and High Frequency," Tesla described many of his discoveries which are only coming into general use today and are being hailed as modern inventions. Among these are the "neon" and other gas-filled lamps, and phosphorescent lamps. Many of the discoveries described are still unutilized, including, as will be seen, the carbon or metallic-button incandescent lamp, requiring but a single wire connection; and still others, which he later discovered, were rich producers of the mysterious X-rays.

The transcript of these lectures runs to 40,000 words. Scores of pieces of apparatus were used and usually several experiments were performed with each. He described "wireless" lamps, glowing glass tubes that required no wire connection for their operation. He described motors which operated on one wire, and "wireless" or "no wire" motors. But perhaps the most important development he described was the sensitive electronic tube— the original of all our modern radio and other electronic tubes— which, he predicted, was the device that would permit receiving wireless telegraph messages across the Atlantic. Of all these discoveries we shall presently have more to say in detail.

It had been Tesla's intention to make a short visit to his early home in Gospic when his lectures were out of the way, but circumstances forced him to make the trip sooner than he expected. Returning to his hotel after delivering the second Paris lecture, he received word that his mother was gravely ill. He rushed to the railroad station, arriving in time to board a train just about to pull out. He telegraphed ahead for special transportation facilities to shorten his trip, and succeeded in reaching Gospic in time to see his mother alive. He arrived in the afternoon and she died that night.

The great anxiety from which Tesla suffered during his sleepless rush from Paris to Gospic caused a patch of hair on the right side of his head to turn white over night. Within a month its jet black color was restored naturally.

Almost immediately after his mother's death, Tesla contracted an illness which incapacitated him for many weeks. When he recovered, he visited his sister Marica, in Plaski, for two weeks. From there he went to Belgrade, the capital of Serbia, where he arrived in May and was received as a national hero.

During the weeks of enforced physical inactivity imposed on him by his illness, Tesla took stock of himself and became thoroughly dissatisfied with the manner in which he had been conducting his life. No human being could feel anything but a pleasurable reaction in response to the adulation that had been heaped upon him during the past two years. Tesla, however, prided himself upon his wisdom in having so designed his life that he would not become a victim of human frailties, but would function far above the normal human level of physical limitations and intellectual activities. Now Tesla saw, in retrospect, that insofar as he had adhered to his superman plan of life, he had succeeded in achieving his goal of producing the works of a superman at a rate which astounded the world. When, however, he submitted to the first blandishments of the lion hunters after his New York lecture in May, 1891, he observed, social activities had cut into his available time and had interfered with his creative activities. He had let the "man magnificent" supersede his "superman," and two years of valuable time had been largely lost. In addition, he had spent that totally unproductive year at the Westinghouse plant. At the close of that period, he had vowed he would never again work for anyone. He now vowed that he would put an end to the vacuous social activities into which he had been inveigled.

It was not easy for Tesla to live up to his good resolutions, for his European trip had greatly enhanced his fame and triumphant celebrations were scheduled on his reappearance in New York. Nevertheless, he rejected all invitations. He returned to the Hotel Gerlach, where he lived a solitary existence. With a pent-up reserve of physical energy owing to his long abstinence from his heavy daily routine of work, he plunged with great vigor into his new program which was to open up new and enchanting realms of scientific wonders.

## SEVEN

THE first public application of Tesla's polyphase alternating-current system was made at the Chicago World's Fair, the Columbian Exposition, which opened in 1893 to celebrate the four-hundredth anniversary of the discovery of America. This was the first world's fair for which electric lighting was a possibility, and the architects availed themselves of the opportunities it afforded for obtaining spectacular effects in illuminating the grounds and buildings at night, as well as for interior lighting during the day. The Westinghouse Electric Company secured the contract for installing all power and lighting equipment at the Fair, and took full advantage of this opportunity to use the Tesla system and demonstrate its great versatility. It supplied all the current used for lighting and power.

While the Chicago World's Fair was in reality a monument to Tesla, he had, in addition, a personal exhibition in which he demonstrated his most recent inventions. One of his exhibits was a spinning egg, made of metal. The egg was shown lying on top of a small velvet-covered circular platform. When Tesla closed a switch the egg stood on its small end and rotated at a high speed as if by magic. The "magic" phase of this feat appealed to a public which, however, grasped little of the explanation that it illustrated the principle of the rotating magnetic field produced by the polyphase alternating currents. In other of his exhibits, glass tubes suspended in space or held in his hands lighted up in an equally "magical" fashion.

But his most spectacular feat was to let 1,000,000 volts pass through his body. This was alternating current of very high frequency as well as high voltage. He had discovered means of producing such currents. Eight years had passed since Edison, attacking high-voltage alternating current as deadly, had refused to become interested in Tesla's polyphase system. Now the Tesla system was providing the electricity for the great world's fair and the Edison direct-current system was ignored. The final gesture of victory was for Tesla to answer Edison's charge that alternating current was deadly by passing the highest voltage of it ever produced through his own body for many minutes without the slightest sign of harm. This bit of showmanship endeared Tesla to the public and brought him a tremendous burst of world-wide fame. Unfortunately, however, it obscured his more important work with polyphase currents.

The next great achievement to be attained by his polyphase system was the harnessing of Niagara Falls.* In 1886 a charter

* Before this was done, and even before the opening of the Chicago fair, the practicability of his system was demonstrated in Europe; but this had been undertaken without his knowledge. A practical test of the transmission of polyphase alternating currents at 30,000 volts was made between a hydroelectric

had been granted for developing power at the Falls. The project made slow progress and was taken over by a New York group which organized the Cataract Construction Company, of which Edward Dean Adams was made president. Mr. Adams' company desired to develop power on the largest scale possible. The total energy supply available in the Falls had been variously estimated from 4,000,000 to 9,000,000 horsepower. Mr. Adams organized the International Niagara Commission for the purpose of ascertaining the best means of harnessing the Falls, and made Lord Kelvin, the famous English scientist, its chairman. A prize of $3,000 was offered for the most practical plan submitted.

Tesla had predicted nearly thirty years before, as a boy, that he would someday harness Niagara Falls. Here was the opportunity. In the meantime, he had made it possible to fulfil his boyhood boast by completing the series of inventions which made it possible to change the hydraulic power of the Falls into electrical energy.

The prize-offer plan adopted by Mr. Adams did not, however, set well with Mr. Westinghouse when he was urged to submit a proposal. He replied, "These people are trying to get one hundred thousand dollars' worth of information for three thousand dollars. When they are ready to talk business we will submit our plans." This adamant attitude of Westinghouse was one handicap for the Tesla alternating-current plan. The second big handicap was the fact that Lord Kelvin had declared himself in favor of the use of direct current.

About twenty plans were submitted in the contest but none of them was accepted by the commission, and no prize was awarded. The big electrical companies, Westinghouse, Edison General Electric and Thomson-Houston, did not submit plans. This took place in 1890.

---

station at Lauffen and the City of Frankfurt, the current being used to furnish electricity at a fair held at the latter city. This installation was built in 1891. The current was used to light incandescent and arc lamps and also to operate a Tesla motor.

Original developers of the Falls planned to use locally the mechanical power provided by water wheels; but the only practical plan was, clearly, the generation of electricity by dynamos driven by water wheels, and the distribution of the current throughout the district. There was a good additional market for it at Buffalo, a large industrial city about twenty-two miles distant. There was always the hope, too, that the current could be transmitted to New York City and serve the rich intervening territory. If direct current were used, its transmission twenty-two miles to Buffalo was totally unfeasible. The Tesla alternating-current system, however, made the transmission to Buffalo extremely practicable and the delivery of the current to New York City a possibility.

In due time the Cataract Construction Company decided that the hydroelectric system was the only feasible one, and proposals and bids were asked on a power system consisting of three generating units, each of 5,000 horsepower, from the Westinghouse Electric Company and the General Electric Company. Each one submitted a proposal to install a Tesla polyphase generating system. The General Electric Company, successor to the Edison General Electric Company, having in the meantime secured a license to use the Tesla patents, proposed to install a three-phase system, and Westinghouse a two-phase system. The first proposal concerned the building of the powerhouse. A second proposal on which bids were asked concerned the transmission line between Niagara Falls and Buffalo and a distribution system in the latter city.

Bids were asked for early in 1893, and in October of that year Mr. Adams announced that the Westinghouse plan for the powerhouse and the General Electric plan for the transmission line were accepted. The latter included a transformation of the two-phase current from the generators into three-phase current to be transmitted to Buffalo. This change indicated the flexibility of the Tesla polyphase system.

Westinghouse completed the powerhouse and in 1895 it stood

ready to deliver 15,000 horsepower; the most gigantic piece of electrical engineering conceived or accomplished up to that time. In 1896 General Electric completed the transmission and distribution system, and electrical power extracted from Niagara Falls, without in any way impairing the beauty of the spectacle they presented, was delivered to industries through the Falls and Buffalo areas. So successful was this installation that the Westinghouse Company installed seven additional generating units, bringing the output to 50,000 horsepower. A second equivalent powerhouse, also using alternating current, was later built by the General Electric Company. Today, the powerhouses at Niagara Falls are linked directly with the electric power system in New York City, all using the Tesla system.

Dr. Charles F. Scott, Professor Emeritus of Electrical Engineering at Yale University, and former president of the American Institute of Electrical Engineers, who was a Westinghouse engineer when that company was developing the Tesla system, in a memorial review of Tesla's accomplishments,* describes the Niagara development and its results:

The simultaneous development of the Niagara project and the Tesla system was a fortuitous coincidence. No adequate method of handling large power was available in 1890; but while the hydraulic tunnel was under construction, the development of polyphase apparatus justified the official decision of May 6, 1893, five years and five days after the issuing of Tesla's patents, to use his system. The polyphase method brought success to the Niagara project; and reciprocally Niagara brought immediate prestige to the new electric system.

Power was delivered in August 1895 to the first customer, the Pittsburgh Reduction Company (now Aluminum Company of America) for producing aluminum by the Hall process, patented in the eventful year 1886. . . .

In 1896 transmission from Niagara Falls to Buffalo, 22 miles, was inaugurated. Compare this gigantic and universal system capable of uniting many power sources in a superpower system, with the multiplicity of Lilliputian "systems" which previously supplied electrical

* Published in *Electrical Engineering*, August 1943, pp. 351–355.

service. As Mr. Adams aptly explained: "Formerly the various kinds of current required by different kinds of lamps and motors were generated locally; by the Niagara-Tesla system only one kind of current is generated, to be transmitted to places of use and then changed to the desired form."

The Niagara demonstration of current for all purposes from large generators led immediately to similar power systems in New York City—for the elevated and street railways and for the subway; for steam railway electrification; and for the Edison systems, either by operating substations for converting alternating current to direct current or by changing completely to A.C. service.

The culminating year 1896 inaugurated two far reaching developments for the extension of polyphase power, one commercial and one engineering. By exchange of patent rights, the General Electric Company obtained license rights under Tesla patents, later made impregnable by nearly a score of court decisions. Also the Parsons turbine, accompanied by its foremost engineer, was transplanted to America and enabled George Westinghouse to bring to fruition by a new method the ideal of his first patent, a "rotary steam engine." The acme of the reciprocating engine came in the early 1900s; a century's development produced the great engines that drove 5,000 to 7,500 kilowatt alternators for New York's elevated and subway. But the rapidly growing steam turbine of different types soon doomed the engine to obsolescence; single units with the capacity of a score of the largest engines are now supplying power to the metropolis. Single powerhouses now supply more power than all of the thousands of central stations and isolated plants of 1890.

Prof. Scott concludes: "The evolution of electric power from the discovery of Faraday in 1831 to the initial great installation of the Tesla polyphase system in 1896 is undoubtedly the most tremendous event in all engineering history."

Lord Kelvin, who had originally favored direct current for Niagara, later conceded, but only after the system was in operation, that alternating current had many more advantages for long-distance distribution systems, and declared, "Tesla has contributed more to electrical science than any man up to his time."

THERE should never have been the slightest shadow of doubt concerning the credit due to Tesla not only for discovering the rotating magnetic field but also for inventing the first practical alternating-current motor, the polyphase system of alternating currents, dynamos for generating them, a variety of motors for converting the currents into power, a system of polyphase transformers for raising and lowering voltages, and economical methods for transmitting electrical power for long distances. Nevertheless, credit for priority has unjustly been given to and taken by others. Tesla succeeded in establishing his claims; but in the meantime, however, damage was done by raising these unfair claims, and to this day the electrical engineering profession, and public service and major electrical industries, have never extended to Tesla the credit to which he is entitled. If they had done so, the name of Tesla would carry at least as much fame as the names Edison and Westinghouse.

Tesla, as we have seen, made his rotating magnetic field invention in 1882, and within two months evolved the complete power system, including all the apparatus which he later patented. In 1883 he described his invention to officials of the Continental Edison Company. In 1884 he demonstrated his motor to the mayor of Strassburg and others. In this same year he described the invention to Thomas A. Edison. In 1885 he sought to have the promoters of the Tesla Arc Light Company develop his system. In 1887 he secured financial backing and built a series of the dynamos and motors which were tested by Prof. Anthony of Cornell University. On October 12, 1887, the first patent applications covering his basic inventions were revealed to the Patent Office. The patents were granted on various dates in the early months of 1888. On May 16, 1888, he presented a demonstration and description of his basic inventions before the American Institute of Electrical Engineers in New York. So much for the record.

The first complication arose when Prof. Galileo Ferraris, a physicist in the University of Turin, presented a paper on *"Rotazioni elettrodynamiche"* (Electrodynamic Rotation) before the Turin Academy in March, 1888. This was six years after Tesla made his discovery, five years after he demonstrated his motor and six months after he had applied for patents on his system. Prof. Ferraris had been carrying on researches in the field of optics. The problem that particularly interested him was polarized light. In this period it was considered necessary to build mechanical models to demonstrate all scientific principles. It was not very difficult to devise models to demonstrate the nature of plane-polarized light, but circularly polarized light presented a more difficult problem.

Prof. Ferraris gave some thought to this problem in 1885, but made no progress until 1888 when he turned to alternating currents for a solution. In that period light was erroneously thought of as a continuously undulating wave in the ether. Prof. Ferraris took the continuously alternating current as an analogue of the plane-polarized light wave. For a mechanical analogue of the circularly polarized light wave he visualized a second train of waves 90 degrees out of step with the first, giving a right-angle vector to the component that should manifest itself by rotation. This paralleled the solution at which Tesla had arrived six years earlier.

In arranging a laboratory demonstration Prof. Ferraris used a copper cylinder suspended on a thread to represent the light waves, and caused two magnetic fields to operate on it at right angles to each other. When the currents were turned on, the cylinder rotated, wound up the thread on which it was suspended and raised itself. This was an excellent model of rotary polarized light waves. The model bore no resemblance to a motor, nor did the Turin scientist have any intention that it should be so considered. It was a laboratory demonstration in optics, using an electrical analogy.

Prof. Ferraris' next experiment mounted the copper cylinder on a shaft and divided each of his two coils into two parts, placing one on either side of the copper cylinder. The device worked up to a speed of 900 revolutions per minute—and beyond this point lost power so rapidly it ceased to operate entirely. He tried iron cylinders but they did not work nearly so well as the copper ones. Prof. Ferraris predicted no future for the device as a power source, but he did predict it would find usefulness as the operating principle for a meter for measuring current.

Prof. Ferraris thus demonstrated that he failed by a wide margin to grasp the principle which Tesla developed. The Italian scientist found that the use of the magnetic iron cylinder interfered with the operations of his device, whereas Tesla, following the correct theory, utilized iron cores for the magnetic field of his motor, used an iron armature, and obtained an efficiency of about 95 per cent in his first motor, which had a rating of about a quarter horsepower. The efficiency of Ferraris' device was less than 25 per cent.

It was Prof. Ferraris' belief that he had performed an important service to science by demonstrating that the rotating magnetic field could *not* be used on any practical basis for producing mechanical power from alternating current. He never deviated from this conclusion, nor did he ever claim that he had anticipated Tesla's discovery of a practical means for utilizing the rotating field for producing power. Knowing that his process was entirely different from Tesla's, he never advanced a claim to independent discovery of the alternating-current motor. He even conceded that Tesla had arrived at his discovery of the rotating magnetic field entirely independently of him, and that Tesla could not in any way have known of his work before publication.

A description of Prof. Ferraris' experiments, however, was published in *The Electrician*, in London, May 25, 1888 (page 86). This was accompanied by the statement:

Whether the apparatus devised by Prof. Ferraris will lead to the discovery of an alternating current motor is a question we do not pretend to prophesy, but as the principle involved may also have other applications, notably in the construction of meters for measuring the supply of electricity . . .

A year before this time Prof. Anthony had already tested Tesla's alternating-current motors in the United States and reported that they attained an order of efficiency equal to that of direct-current motors; and Tesla's U. S. patents had been publicly announced several months previously.

It was obvious that the editors of this London publication were not keeping up to date on developments in the United States.

Tesla responded quickly, informing the editors of their oversight and submitting an article describing his motors and the results obtained with them.

No great enthusiasm was exhibited by the editors of *The Electrician*. They receded to only the least possible extent from their stand in favor of Ferraris by publishing an editorial note: *

Our issue of the 25th of May contained an abstract of a paper by Prof. Galileo Ferraris describing a method of producing a revolving resultant magnetic field by means of a pair of coils with the axes at right angles and traversed by alternating currents, and we drew attention to the possibility that the principle of the apparatus might be applied to the construction of an alternating current motor. The paper by Mr. Nikola Tesla, which appears in our columns this week, contains a description of such a motor, founded on exactly the same principle.

No attention was drawn to the fact that Ferraris had reached the conclusion that the principle could never be used for making a practical motor, whereas Tesla had produced such a motor.

This attitude toward the American development did not disappear from the London engineering journals. Later the *Electrical Review* † published an editorial which opened with the statement:

For several years past, from the days of Prof. Ferraris' investigations, which were followed by those of Tesla and Zipernowski and a host of imitators, we have periodically heard of the question of alternating current motors being solved.

* Vol. XX, p. 165, June 15, 1888.
† London: Vol. XXVIII, p. 291, March 6, 1891.

At this time the Westinghouse Company was already commercially exploiting the successful and practical Tesla polyphase system in the United States. Not one word of credit to Tesla appeared in the London engineering press.

A letter of protest dated March 17, 1891, was forwarded by Tesla, and this was published some weeks later (p. 446) by the *Review*. He said in part:

> In all civilized countries patents have been obtained almost without a single reference to anything which would have in the least degree rendered questionable the novelty of the invention. The first published essay—an account of some laboratory experiments by Prof. Ferraris—was published in Italy six or seven months after the date of filing my application for the foundation patents. . . . Yet in your issue of March 6, I read: "For several years past, from the days of Prof. Ferraris' investigations, which were followed by those of Tesla and Zipernowski and a host of imitators, we have periodically heard of the question of alternating current motors being solved."
>
> No one can say that I have not been free in acknowledging the merit of Prof. Ferraris, and I hope that my statement of facts will not be misinterpreted. Even if Prof. Ferraris' essay would have anticipated the date of filing of my application, yet, in the opinion of all fair minded men, I would have been entitled to the credit of having been the first to produce a practical motor; for Prof. Ferraris denies in his essay the value of the invention for the transmission of power. . . .
>
> Thus in the most essential features of the system—the generators with two or three currents of differing phase, the three wire system, the closed coil armature, the motors with direct current in the field, etc.,—I would stand alone, even had Prof. Ferraris' essay been published many years ago. . . .
>
> Most of these facts, if not all, are perfectly well known in England; yet according to some papers, one of the leading English electricians does not hesitate to say that I have worked in the direction indicated by Prof. Ferraris, and in your issue above referred to it seems I am called an imitator.
>
> Now, I ask you where is that well known English fairness. I am a pioneer and I am called an imitator. I am not an imitator. I produce original work or none at all.

This letter was published; but the *Electrical Review* neither expressed regret for the misstatement nor extended recognition to Tesla.

Charles Proteus Steinmetz, later to achieve fame as the electrical wizard of the General Electric Company, came to the support of Tesla. In a paper presented before the American Institute of Electrical Engineers, he said: * "Ferraris built only a little toy, and his magnetic circuits, so far as I know, were completed in air, not in iron, though that hardly makes any difference."

Other American engineers likewise rallied to Tesla's support.

An industrial exposition, as already mentioned, was held at Frankfurt, Germany, in 1891. The United States Navy sent Carl Hering, an electrical engineer who had done much writing for technical journals, as observer to report on any developments that would be of interest to the Navy. Hering, unfortunately, had not informed himself of the inventions embodied in the Tesla patents before going abroad.

The outstanding new development at the Frankfurt exposition was the first public application of Tesla's system. The grounds and building were lighted by electricity brought to the city by a long-distance transmission line over which electricity was carried from the hydroelectric station at Lauffen by three-phase alternating current carried at 30,000 volts. There was exhibited a two-horsepower motor operated by the three-phase current.

Hering recognized the significance of the new development, and sent back enthusiastic reports describing it as of German origin. In his article in the *Electrical World* (N.Y.), he waxed enthusiastic about the work of Dolivo Dobrowolsky in designing the three-phase motor and its associated system, hailing it as an outstanding scientific discovery and of tremendous commercial importance. The impression was given that all other inventors had missed the main point, and that Dobrowolsky had

* Transactions, A.I.E.E., Vol. VIII, p. 591, 1891.

achieved the grand broad accomplishment that would set the pace for future power developments. Nor was Hering the only one to whom this impression was communicated.

Ludwig Gutman, an American electrical engineer, a delegate to the Frankfurt Electrical Congress, in a paper on "The Inventor of the Rotary Field System." delivered before that body,* slammed back at Dobrowolsky. He stated:

As we have enjoyed in America several years' experience with this system represented by the Tesla motors I must oppose the assertion lately made by Herr von Dobrowolsky at a meeting of the Electrotechnische Gesellschaft held here in Frankfurt. The gentleman said; "I believe I am able to assert that the motor problem for large and small works has been by this completely solved." This assertion goes most likely too far. The problem was already solved, theoretically and electrically, in 1889.

Dobrowolsky, in a paper published in the *Electrotechnische Zeitschrift* (p. 149–150; 1891), reduced his claim to that of having produced the first practical alternating-current motor; and he asserted that in the Tesla two-phase motor there were field pulsations amounting to 40 per cent, while in his three-phase motor, in operation at the Frankfurt exposition, these were greatly reduced.

Even this reduced claim of Dobrowolsky's was quickly smashed. It drew fire from an American and an English source, and also from the chief engineer of the project of which his motor was a part.

Dr. Michael I. Pupin, of the Department of Engineering, Columbia University, analyzing Dobrowolsky's claim,† demonstrated that he had failed to comprehend the basic principles of the Tesla system, and that the three-phase system which he claimed as his own was included in Tesla's inventions.

C. E. L. Brown, the engineer in charge of the pioneering Lauffen-Frankfurt 30,000-volt transmission system and its three-

---

* *Electrical World*, N.Y.: Oct. 17, 1891.
† *Ibid.*, Dec. 26, 1891.

phase generating system, including the Dobrowolsky motor, settled definitely and completely the question of credit for the whole system. In a letter published in the *Electrical World* (Nov. 7, 1891), he concluded with the statement: "The three phase current as applied at Frankfurt is due to the labors of Mr. Tesla and will be found clearly specified in his patents."

Mr. Brown wrote letters to other technical publications to this same effect, and in them criticized Mr. Hering for failing to give Tesla his due credit, and for diverting it to Dobrowolsky.

These criticisms finally brought a response from Mr. Hering. This appeared in the *Electrical World*, Feb. 6, 1892:

As Mr. C. E. L. Brown, in communications to the Electrical World and other journals, seems determined to insist that I have neglected the work of Mr. Tesla on rotary current I wish to state there is no one more willing than myself to give Mr. Tesla due credit for his work, and I have always considered him to be an original inventor of the rotary field system and first to reduce it to practice, and I believe I so stated in my articles. If I have at any time failed to give him credit for the extent to which he developed it, it has been because Mr. Tesla has been too modest (or perhaps prudent) to let the world know what he has accomplished. When the articles which have caused this discussion were being written Mr. Tesla's patents were not accessible to me. Just where Mr. Dobrowolsky's improvements begin I have not been able to ascertain. . . .

Dobrowolsky, though he may have been an independent inventor, admits Tesla's work is prior to his. . . . The modesty of both of these gentlemen would, I feel sure, lead to a clear understanding. Regarding the subject of priority it may be of interest here to say that in a conversation with Prof. Ferraris last summer that gentleman told me with very becoming modesty that, although he had experimented with the rotary field several years before Tesla's work was published he did not think it was possible that Tesla could have known of his work and he therefore believed Tesla invented it entirely independently. He also stated that Tesla developed it much further than he (Ferraris) did.

Thus the scientists and engineers in the United States, Germany and Italy gave Tesla clear and unquestioned credit for being the sole inventor of the magnificent polyphase electrical system in all of its details. French and British journals then fell in line.

Thus, by 1892, there was universal acclaim for Tesla as the unquestioned inventor of the alternating-current motor and the polyphase system in engineering circles. There was none, therefore, to dispute his claim or to seek to rob him of credit when his fame reached the public through the operation of his system at the World's Fair in Chicago in 1893, and later when his system made possible the harnessing of Niagara Falls.

In due time, however, there came many who claimed to have made improvements on Tesla's inventions; and widespread efforts were made to exploit these "improvements." The Westinghouse Company, now owners of the Tesla patents, undertook to defend the patents and to prosecute infringers. As a result about twenty suits were carried to the courts, and, in every one of them, decisions gave a decisive victory to Tesla.

A sample of the sweeping decisions that were handed down is that of Judge Townsend in the United States Circuit Court of Connecticut in September, 1900, when, passing judgment on the first group of basic patents, he said in part:

It remained to the genius of Tesla to capture the unruly, unrestrained and hitherto opposing elements in the field of nature and art and to harness them to draw the machines of man. It was he who first showed how to transform the toy of Arago into an engine of power; the "laboratory experiment" of Bailey into a practically successful motor; the indicator into a driver; he first conceived the idea that the very impediments of reversal in direction, the contra-indications of alternations might be transformed into power-producing rotations, a whirling field of force.

What others looked upon as only invincible barriers, impassable currents and contradictory forces he seized, and by harmonizing their directions utilized in practical motors in distant cities the power of Niagara.

The resentments and antagonisms engendered by the unvarying series of successful decisions caused individuals who were adversely affected to vent their antagonisms on Tesla although he had not in ten years held any personal interests in the patents.

The situation that developed is well described by B. A. Behrend, later vice-president of the American Institute of Electrical Engineers: *

It is a peculiar trait of ignorant men to go always from one extreme to another, and those who were once the blind admirers of Mr. Tesla, exalting him to an extent which can be likened only to the infatuated praise bestowed on victims of popular admiration, are now eagerly engaged in his derision. There is something deeply melancholy in the prospect, and I can never think of Nikola Tesla without warming up to my subject and condemning the injustice and ingratitude which he has received alike at the hands of the public and of the engineering profession.

With the scientific and engineering worlds, and the courts, extending to him a clear title to the honor of being the great pioneer discoverer and inventor of the principles and machines that created the modern electrical system, Tesla stands without a rival as the genius who gave the world the electrical power age that made our mass-production industrial system possible. The name Tesla should, therefore, in all right and justice, be the most famous name in the engineering world today.

* *Western Electrician*, Sept., 1907.

SECOND PART

*FORTUNE AND FAME*

# EIGHT

RETURNING to his laboratory in March, 1893, after his European and American lectures, Tesla banished all social activities from his life program, and, bursting with energy, pitched headlong into experimental work in connection with his wireless system. He made repeated experiments in working out the refinement of his principle of tuning circuits to resonance with each other. He built more than one hundred coils covering a wide range of electrical tuning characteristics. He also built numerous oscillators for producing high-frequency currents, and condensers and inductances for tuning both sending and receiving coils to any desired frequency or wavelength.

He demonstrated that he could cause any one of hundreds of coils to respond selectively and powerfully to its particular wavelength emitted by an oscillator while all others remained inert; but he discovered that tuned electrical coils have, to a further extent, the same properties as tuned musical string, in that they vibrate not only to the fundamental note but also to a wide range of upper, and particularly lower, harmonics. This characteristic could be usefully employed in connection with the design of sending- and receiving-station antennas, but it militated against the sharp, exclusive response tuning of coils. At close range, and with the powerful currents Tesla used in his laboratory, the harmonics were a handicap—yet when greater distance separated sending and receiving coils, this trouble became a minor one.

It became obvious to Tesla that it was going to be difficult to arrange an early demonstration of his worldwide system of intelligence and of power, so he planned a compromise system in which he would use a smaller central transmitter and smaller relay stations at certain distances.

In an interview with Arthur Brisbane, the famous editor, Tesla announced in *The World* of July 22, 1894, the certainty of his plans. He said:

You would think me a dreamer and very far gone if I should tell you what I really hope for. But I can tell you that I look forward with absolute confidence to sending messages through the earth without any wires. I have also great hopes of transmitting electric force in the same way without waste. Concerning the transmission of messages through the earth I have no hesitation in predicting success. I must first ascertain exactly how many vibrations to the second are caused by disturbing the mass of electricity which the earth contains. My machine for transmitting must vibrate as often to put itself in accord with the electricity in the earth.

During the following winter he designed and built his transmitting station and a receiving station for this purpose. It worked well within the close range of the laboratory and between points in the city. Like the artist who is never willing to declare a picture finished but must continue to apply an unending series of slight improvements, Tesla continued to add refinements so that he would be assured of a perfect test in the spring, when he planned to take his receiving set up the Hudson River on a small boat to test its response at extended distances.

With Tesla, as with Cæsar, though, came tragedy on the Ides of March. For Tesla it was the unlucky 13th of March, 1895, when fire broke out during the night in the lower part of the building in which his laboratory was located and swept through the entire structure. The two floors on which his equipment was located dropped to the basement, their entire contents destroyed. Not a single article was saved. The major portion of Tesla's fortune was invested in the apparatus in that building. He carried no insurance on it. The loss was total.

The monetary loss was the least important factor in the shock which Tesla sustained. The apparatus and the countless experiments in scores of subjects with which they were associated were part of Tesla's self. His work of a lifetime was swept away. All of his records, papers, mementos, his famous World's Fair exhibit were gone. His laboratory, in which he had demonstrated his wonders to the élite and intelligentsia of New York, to the most famous men and women of the country and the world, was no more. And this tragedy had come just when he was ready to make his first distance demonstration of his wireless system.

Tesla was in a tough spot financially. The laboratory was the property of the Tesla Electric Company, owned by Tesla and A. K. Brown who had, with an associate, put up the funds to finance Tesla's demonstration of his polyphase alternating-current system prior to its sale to Westinghouse for $1,000,000. Some of that money was divided as cash among the associates, as stated; and the remainder had gone into the laboratory for further developments. The resources of the company were now wiped out and Tesla's individual resources were almost at the vanishing point. He was receiving some patent royalties from Germany on his polyphase motors and dynamos. This income would be adequate to take care of his living expenses but not sufficient to enable him to maintain an experimental laboratory.

Mr. Adams, active head of the Morgan group that had developed the hydroelectric station at Niagara Falls, using Tesla's polyphase system, now came to the inventor's rescue. He proposed and arranged for the formation of a new company which would finance the continuation of Tesla's experiments, and he offered to subscribe one hundred thousand dollars of the proposed half-million dollars of capital stock of the company.

With this support Tesla proceeded to set up a new laboratory. He secured quarters at 46 East Houston Street, and started operations there in July, 1895, four months after his South Fifth Avenue laboratory had been destroyed.

Adams paid forty thousand dollars as the first installment of his subscription. He took an active personal interest in Tesla's work, and spent a great deal of time in the laboratory. Knowing from the successful operation of the Niagara Falls plant that Tesla, technically, was extremely practical, Adams was deeply impressed by the plans for wireless transmission of intelligence and of power. He declared he was willing to go still further than his original plan of financial support, and he proposed that the plan include the taking in of his son as an active partner in Tesla's work.

Such an arrangement would amount to an alliance for Tesla with the powerful Morgan financial group. It was the support of J. P. Morgan that gave financial guidance to the formation of the General Electric Company and made possible the building of the Waterside Station, the first big Edison powerhouse in New York, and it was a Morgan group that, by making possible the development of Niagara, had given the Tesla system a tremendous boost. The prestige that would come from a Morgan association would probably be even more potent than the actual monetary aid involved. With this alliance Tesla's financial future was assured. There would come to his aid, through it, the support of the world's greatest organizational genius and practical promotion powers. The tragedy of the fire that brought about this situation could yet prove a great blessing.

Tesla made his decision. What influenced him to reach the determination that guided him, no one ever learned. He rejected Mr. Adams' offer! From a practical point of view there is no way of explaining his action. But no one could ever successfully demonstrate that Tesla was practical in a commercial and financial sense.

With the forty thousand dollars that Adams subscribed, Tesla was able to keep actively engaged in research for about three years. He probably could have secured subscriptions of many times that amount if he had been willing to put forth even a slight effort in that direction, but he was interested mainly in getting his experiments well under way rather than worrying about future financial needs. He had full faith that the future would bring him many millions of dollars as a token of the many billions of value he would give it through his inventions.

It took Tesla about a year to get his laboratory equipped and to build an array of experimental apparatus. Almost nothing that he used could be purchased in the market; everything had to be specially made by his workmen under his direction. In the spring of 1897 he was ready to make, on his wireless transmitter and receiver, the distance tests which had been interrupted by the fire two years before.

The success of these tests were announced by Tesla in an interview with a representative of the *Electrical Review* which was published in the issue of July 9, 1897, of that journal. It stated:

Nearly every telegraphic inventor has for years dreamed in his waking hours of the possibility of communicating without wires. From time to time there has appeared in the technical journals a reference to the experiments showing the almost universal belief among electricians that, some day, wires will be done away with. Experiments have been made attempting to prove the possibilities, but it has remained for Mr. Nikola Tesla to advance a theory, and experimentally prove it, that wireless communication is a possibility and by no means a distant possibility. Indeed, after six years of careful and conscientious work, Mr. Tesla has arrived at a stage where some insight into the future is possible.

A representative of the Electrical Review receives the assurance personally from Mr. Tesla who, by the way, is nothing if not conservative, that electrical communication without wires is an accomplished fact and that the method employed and the principles involved have nothing in them to prevent messages being transmitted and intelligibly received between distant points. Already he has constructed both a transmitting apparatus and an electrical receiver which at distant points is sensitive to the signals of the transmitter, regardless of earth currents or points of the compass. And this has been done with a surprisingly small expenditure of energy.

Naturally, Mr. Tesla is averse to explaining all details of his invention, but allows it to be understood that he avails himself of what, for the present, may be termed the electrostatic equilibrium; that if this be disturbed at any point on the earth the disturbance can with proper apparatus be distinguished at a distant point and thus the means of signalling and reading signals becomes practicable once the concrete instruments are available. Mr. Tesla announced his belief in the possibilities, but he did so after having satisfied himself by actual test of apparatus designed by him. Much work has yet to be done, and he has since then given close attention and study to the problem.

Details are not yet available, for obvious reasons, and we now merely chronicle Mr. Tesla's statement that he has really accomplished wireless communication over reasonably long distances with small expenditure of energy and has only to perfect apparatus to go to any extent. Morse's 40 mile experiment in the old days was on a far less certain basis than the wireless possibilities of today.

Tesla's work with high frequency and high potential currents has been notable. As long ago as 1891 he foretold the present results, both as to vacuum tube lighting and intercommunication without wires. The former has in his hands assumed a condition capable of a public demonstration of the phenomena of the electrostatic molecular forces. Numberless experiments were carried out, and from what then was a startling frequency of 10,000 per second Mr. Tesla has advanced to what now is a moderate rate at 2,000,000 oscillations per second.

This announcement recorded the birth of modern radio—radio as it is in use today—born on a boat traveling up the Hudson River, carrying the receiving set twenty-five miles from the Houston Street laboratory, a distance which was a small fraction of the range of the set but enough to demonstrate its capabilities. Such an accomplishment was worthy of a flamboyant smash announcement instead of Tesla's very modest statement and the even more conservative manner in which the *Electrical Review* treated the news. Tesla had to protect not only his patent rights, which would be jeopardized by premature disclosure, but also had to be on guard against invention invaders and patent pirates, with whom he had previously had unpleasant experiences. The *Electrical Review*, naturally enough, was fearful of the consequences of "sticking its neck out" by too enthusiastic a reception before full details were available.

The fundamental patents on Tesla's system were issued on September 2, 1897, just two months after his announcement. They are numbered 645,576 and 649,621. In these patents he describes all the fundamental features of the radio broadcasting and receiving circuits in use today. Once patent protection was secured, Tesla did not long delay in letting the public in on his discoveries. His presentation took the form of a spectacular demonstration at Madison Square Garden.

WIRELESS transmission of intelligence is a modern satisfaction of one of the oldest cravings of man, who has always sought the annihilation of distance by communication through space without material linkage over the intervening expanse. Early experimenters with the telephone, particularly, were enthusiastic seekers of a method of wireless electrical communication that would convey the voice through space in the manner in which the air conducted sound. David Edward Hughes had noted, in 1879, that when an electric spark was produced anywhere in his house he heard a noise in his telephone receiver. He traced the effect to the action of the carbon granules in contact with a metal disk in his telephone transmitter which acted as a detector of the space waves by sticking together slightly, reducing the resistance of the mass, and producing a click in the receiver.

Prof. A. E. Dolbear, of Tufts College, amplified this observation and set up, in 1882, a demonstration set using the principle but eliminating the telephone set. He used a spark coil for creating waves and a mass of carbon granules for detecting them. This is exactly the "wireless" system which Marconi "discovered" fourteen years later.

Edison, engaged by the Western Union Telegraph Company to break the monopoly which Bell held by his invention of the telephone, had succeeded, in 1885, in sending a message from a moving train by "wireless." A wire strung on the train paralleling a telegraph wire strung on poles along the track made it possible to bridge the intervening few feet by an inductive effect—the same effect which causes annoyance by creating "cross talk," or a mixing of conversations over two telephone circuits located close to each other. W. M. Preece, in England, made a similar experiment about the same time. The extremely short distances over which such systems worked prevented them from having any practical usefulness.

An entirely different type of wireless communication had been developed by Alexander Graham Bell in 1880 and 1881. This was given the name radiophone, but Bell insisted on calling it the photophone. The photophone transmitted the voice over a beam of light. The transmitter consisted of a very thin glass or mica mirror, which could be vibrated by the voice. This reflected a beam of light, usually sunlight, to a distant receiving device. The simple receiver consisted of a chemist's test tube, into which a selected material was placed. The top of the tube was closed by a cork through which two small rubber tubes were inserted, the other ends being placed in the ears. A very great variety of materials could be placed in the test tube as detectors. When the beam of light, vibrated by the voice, impinged on the material in the tube, an absorption of heat took place which set the air in the tube in vibration, thus reproducing the voice that was carried by the light beam. Bell also used selenium as a detector. It responded to the visible rays and produced an electrical effect. The experiments, obviously, were of little practical value as the basis for a system of wireless communication.

Michael Faraday, in London, had described in 1845 his theory of the relationship between light and the electromagnetic lines of force; and in 1862 James Clerk Maxwell published an analysis of Faraday's work which gave a mathematical basis for the theory that light waves were electromagnetic in nature, and that it was possible for such waves to exist very much shorter and very much longer than the known wavelength of visible light. This was a challenge to scientists to prove the existence of such waves.

Prof. Heinrich Hertz, at Bonn, Germany, from 1886 to 1888, undertook the search for the waves longer than light or heat. He produced them by the spark discharge of an induction coil and recaptured them from space, at short distances, in the form of a tiny spark that jumped the gap in a slotted ring of wire. Sir Oliver Lodge, in England, was simultaneously seeking to measure equally small electrical waves in wire circuits.

This, then, had been the situation in the scientific world when Tesla began his work in 1889. The plan for wireless communication which he presented in 1892 and 1893, as will be described in a moment, shows how his magnificent concept and tremendously advanced knowledge towered mountain high over all contemporaries.

When Tesla left the Westinghouse plant in the fall of 1889, he had immediately turned to the next phase of his development of the alternating-current field—a new system of distributing energy by means of high-frequency alternating currents which would be a far more magnificent discovery than his polyphase system. Within the next two years he had explored the principles by which energy could be distributed broadcast without the use of wires, and these he had demonstrated with powerful coils in his laboratory. The distribution of intelligence, later called "wireless," was but a single phase of the larger project.

Tesla described, in 1892, the first electronic tube designed for use as a detector in a radio system, and demonstrated its characteristics in his lectures in London and Paris in February and March of that year. (The tube, however, had been developed in 1890.) He described in February and March of the following year, 1893, his system of radio broadcasting, presenting its principles in detail, in lectures before the Franklin Institute in Philadelphia and at the convention of the National Electric Light Association held in St. Louis.

Tesla's electronic tube, his 1890 invention, was the ancestor of the detecting and amplifying tubes in use today. His demonstration of this tube is a matter of record in the archives of four societies before which he exhibited it in February and March of 1892—the Institute of Electrical Engineers and the Royal Society of London and the Physical Society of France and the International Society of Electrical Engineers in Paris. He stated in these lectures:

> If there is any motion which is measurable going on in space, such a brush ought to reveal it. It is, so to speak, a beam of light, frictionless, devoid of inertia.
>
> I think it may find practical applications in telegraphy. With such a brush it would be possible to send dispatches across the Atlantic, for instance, with any speed, since its sensitiveness may be so great the slightest changes will affect it.

The "brush" in Tesla's tube was a beam of electrons. The electron, however, had not yet been discovered. Nevertheless, Tesla gave an accurate description of its nature, demonstrating the remarkable accuracy of his interpretation of strange phenomena. So sensitive was this electronic beam that a small horseshoe magnet an inch wide at a distance of six feet caused movement of the electron beam in either direction, depending on the position in which the magnet was held.

If anyone approached the tube from a distance of many feet the beam, or brush, would swing to the opposite side of the tube. If one walked around the tube even at a distance of ten feet, the beam would move likewise, keeping its center end always pointed at the moving object. The slightest movement of a finger, or even the tensing of muscle, would bring a swinging response from the beam.

In the same 1892 lecture in which he described this first electronic tube, Tesla demonstrated lamps which were lighted without wire connections (wireless light) and also a motor which operated without wire connections to the energizing coils (wireless power); and he had again presented these developments at his exhibition at the Chicago Columbian Exposition early in 1893.

It was with all this experience behind him, giving him full assurance that his system was entirely practical and operative, that Tesla presented at the Franklin Institute and at the convention of the National Electric Light Association in February and March, 1893, a very cautious and conservative statement concerning his plan. Even at these 1893 lectures, Tesla could have staged a demonstration of wireless transmission of intelligence by placing one of his resonant coils, surmounted by one of his electronic "brush" tubes, or one of his low-pressure air lamps, in the lecture hall and causing it to respond to signals sent out by an energized coil of similar wavelength but located at a considerable distance from the building. The experiment was a standard procedure in his laboratory.

This, however, would be a purely local effect, whereas his radio transmission system was one planned on a world-wide basis, requiring much more powerful apparatus than he had thus far built. To pass off a purely local effect as a demonstration of a world-wide system, even though the observed results would have been identical, would have been a case of intellectual dishonesty to which Tesla would not stoop; yet this demonstration of wireless would have been more spectacular and powerful than any staged by any other inventor in more than a half-dozen years following.

Describing his world-wide system at the 1893 National Electric Light Association meeting, he said:

In connection with resonance effects and the problems of transmission of energy over a single conductor, which was previously considered, I would say a few words on a subject which constantly fills my thoughts, and which concerns the welfare of all. I mean the transmission of intelligible signals, or, perhaps, even power, to any distance without the use of wires. I am becoming more convinced of the practicability of the scheme; and though I know full well that the great majority of scientific men will not believe that such results can be practically and immediately realized, yet I think that all consider the developments in recent years by a number of workers to have been such as to encourage thought and experiment in this direction. My conviction has grown so strong that I no longer look upon the plan of energy or intelligence transmission as a mere theoretical possibility, but as a serious problem in electrical engineering, which must be carried out some day.

The idea of transmitting intelligence without wires is the natural outcome of the most recent results of electrical investigations. Some enthusiasts have expressed their belief that telephony to any distance by induction through air is possible. I cannot stretch my imagination so far, but I do firmly believe that it is practical to disturb, by means of powerful machines, the electrostatic conditions of the earth, and thus transmit intelligible signals, and, perhaps, power. In fact, what is there against carrying out such a scheme?

We now know that electrical vibrations may be transmitted through a single conductor. Why then not try to avail ourselves of the earth for this purpose? We need not be frightened by the idea of distance. To the weary wanderer counting the mileposts, the earth may appear very large; but to the happiest of all men, the astronomer, who gazes at the heavens, and by their standards judges the magnitude of our globe, it appears very small. And so I think it must seem to the electrician; for when he considers the speed with which an electrical disturbance is propagated through the earth, all his ideas of distance must completely vanish.

A point of great importance would be first to know what is the capacity of the earth, and what charge does it contain if electrified. Though we have no positive evidence of a charged body existing in space without other oppositely electrified bodies being near, there is a fair probability that the earth is such a body, for by whatever process it was separated—and this is the accepted view of its origin —it must have retained a charge, as occurs in all processes of mechanical separation. . . .

If we can ever ascertain at what period the earth's charge, when disturbed, oscillates, with respect to an oppositely charged system or known circuit, we shall know a fact possibly of the greatest importance to the welfare of the human race. I propose to seek for the period by means of an electrical oscillator or a source õf alternating currents.

One of the terminals of this source would be connected to the earth, as, for instance, to the city water mains, the other to an insulated body of large surface. It is possible that the outer conducting air strata or free space contains an opposite charge, and that, together with the earth, they form a condenser of large capacity. In such case the period of vibration may be very low and an alternating dynamo machine might serve for the purpose of the experiment. I would then transform the current to a potential as high as it would be found possible, and connect the ends of the high tension secondary to the ground and to the insulated body. By varying the frequency of the currents and carefully observing the potential of the insulated body, and watching for the disturbance at various neighboring points of the earth's surface, resonance might be detected.

Should, as the majority of scientific men in all probability believe, the period be extremely small, then a dynamo machine would not do, and a proper electrical oscillator would have to be produced, and perhaps it might not be possible to obtain such rapid vibrations. But whether this be possible or not, and whether the earth contains a charge or not, and whatever may be its period of vibration, it is certainly possible—for of this we have daily evidence—to produce some electrical disturbance sufficiently powerful to be perceptible by suitable instruments at any point on the earth's surface. . . .

Theoretically, then, it could not require a great amount of energy to produce a disturbance perceptible at a great distance, or even all over the surface of the globe. Now, it is quite certain that at any point within a certain radius of the sources, a properly adjusted self induction and capacity device can be set in action by resonance. But not only this can be done, but another source, s 1, similar to s, or any number of such sources, can be set to work in synchronism with the latter, and the vibration thus intensified and spread over a large area, or a flow of electricity produced to or from source s 1, if the same or of opposite phase to the source s.

I think that, beyond doubt, it is possible to operate electrical devices in a city, through the ground or pipe system, by resonance from an electrical oscillator located at a central point. But the practical solution of this problem would be of incomparably smaller benefit to man than the realization of the scheme of transmitting intelligence, or, perhaps, power, to any distance through the earth or environing medium. If this is at all possible, distance does not mean anything. Proper apparatus must first be produced, by means of which the problem can be attacked, and I have devoted much thought to this subject. I am firmly convinced it can be done, and I hope we shall live to see it done.

The lecture before the Franklin Institute contained a similar statement. An additional paragraph from it can be quoted:

If by means of powerful machinery, rapid variations of the earth's potential were produced, a grounded wire reaching up to some height would be traversed by a current which could be increased by connecting the free end of the wire to a body of some size. . . . . The experiment, which would be of great scientific interest, would probably best succeed on a ship at sea. In this manner, even if it were not possible to operate machinery, intelligence might be transmitted quite certainly.

Tesla thus presented in these lectures the principles which he had learned in his laboratory experiments, during the previous three years, were necessary for successful wireless communication.

Several fundamental requirements were presented which will be understood by any non-technical person who has had even slight experience with radio receiving sets: 1. An antenna, or aerial wire; 2. A ground connection; 3. An aerial-ground circuit containing inductance and capacity; 4. Adjustable inductance and capacity (for tuning); 5. Sending and receiving sets tuned to resonance with each other; and 6. Electronic tube detectors. He had still earlier invented a loud speaker.

These embody the fundamental principles of radio, and are used in every sending and receiving set today.

RADIO as it exists today is, therefore, the product of the genius of Nikola Tesla. He is the original inventor of the system as a whole and of all the principal electrical components. The man who, next to Tesla, is entitled to the greatest amount of credit is Sir Oliver Lodge, the great English scientist. Even Lodge, however, failed to grasp the fundamental picture that Tesla presented.

Lodge, early in 1894, had put a Hertz spark gap in a copper cylinder open at one end; and in this way he produced a beam of ultra-short-wave oscillations which could be transmitted in any direction. He did the same for the receiving set. Since the incoming waves could be received from only one direction, this receiving set was able to locate the direction from which the transmitted waves came. With this set he completely anticipated Marconi by two years. In the summer of that year, in a demonstration before the British Association for the Advancement of Science at Oxford, he sent Morse signals, with an improved set, between two buildings separated by several hundred feet.

It is little wonder, then, that Marconi, who started his studies of wireless in 1895, created no stir in the scientific circles in England when he came from Italy to London in 1896 with a wireless set that in every essential feature was the same as that demonstrated by Lodge in 1894. He used a parabolic reflector, so his set was little more than an electrical searchlight. He did, however, bring an alternative feature to replace the parabolic beam reflector. This was a ground connection and antenna, or aerial wire, for both sending and receiving set. This was exactly what Tesla had described in his plan published three years before.

When Hertz made his experiments to demonstrate the identical nature of light and longer electromagnetic waves, he intentionally sought to use the shortest waves it was practicable to produce. They were measured in inches—much less than a yard long. They were entirely satisfactory for his experiment. When the wireless experimenters copied his methods they took over the short-wave plan without ever asking a question as to what wavelength should be used for wireless communication; the thought seems not to have dawned on them that there were other wavelengths that could be produced and used—all except Tesla.

Tesla took the trouble, with the spirit of a real scientist, to repeat exactly the experiments of Hertz; and he published his results, stating that he found a number of important differences and calling attention to the inadequacies of Hertz's experimental methods.

Having experimented with a wide gamut of wavelengths of high-frequency currents and studied the properties of each section of the spectrum, he knew that the short wavelengths were totally unsuitable for communication purposes. He knew that the useful wavelengths ranged from 100 meters to many thousands of meters. He knew that the combination of induction coil and Hertz ball-type spark-gap oscillator could never have any practical usefulness in producing the kind of electrical pulsations required. Even with the highly efficient apparatus available today, scientists have been unable to use in communication (except for special purposes) the ultra-short waves which Tesla in his wisdom condemned and Marconi, owing to his inexperience, tried to use.

The history of the succeeding years in wireless is the story of the failure of the short waves of Lodge and Marconi and their followers, and the shifting over to the longer waves described by Tesla; and the dropping of their crash method of signaling and its replacement by the refined and highly efficient method of tuning to each other the sending and receiving stations by the methods discovered by Tesla; and adoption of Tesla's continuous waves.

In addition, these groping workers saw in wireless only a point-to-point or station-to-station method of signaling. None of them foresaw the broadcasting system which Tesla described in 1893. The system invented and discovered by Tesla is the one in use today; but who ever heard anyone giving Tesla the slightest credit?

# NINE

TESLA was prolific in opening up vast new empires of knowledge. He showered his discoveries on the world at such a rapid rate and in such a nonchalant manner that he seems to have benumbed the minds of the scientists of his age. He was too busy to spend time developing the technical or commercial applications of each new discovery—there were too many other new and important revelations within his vision that must be brought to light. Discoveries were not happenstance events to him. He visualized them far in advance of their unfolding in the laboratory. He had a definite program of pioneering research in virgin fields of investigation; and when this was accomplished he would, he felt, have a long lifetime still ahead of him in which he could return to the practical utilization of those already revealed.

Meanwhile, he had found a whole new world of interesting effects in the discharges produced by his coils when energized with the currents of extremely high frequency. He built larger and larger coils and experimented with a variety of shapes as constructions. From the common cylindrical type of coil he developed the cone-shaped coil, and this development he carried still further by designing the flat helix, or pancake-shaped coil.

The extremely high-frequency currents furnished a mathematical paradise in which Tesla could develop his equations to his heart's content. Through his mathematical abilities and his strange power of visualization he could frequently make, very quickly, whole series of discoveries that it took a long time to catch up with in actual laboratory constructions. This was true of the phenomena of resonance, or tuned circuits.

Because of their relatively short wavelength, it was comparatively easy to build condensers for tuning the circuits. When a circuit is tuned the electric current that flows in it oscillates rhythmically, just as does a musical string which, when struck or plucked, vibrates and builds up loops of even lengths with motionless points between them. There may be only one of these loops, or there can be many.

Tesla did not invent the idea of electrical resonance. It was inherent in the mathematical description of the condenser discharge as developed by Lord Kelvin, and in the physical nature of alternating currents; but Tesla changed it from a buried mathematical equation to sparking physical reality. It is the analogy of acoustical resonance which is a natural property of matter. However, there were no practical circuits in which resonance could manifest itself until Tesla developed alternating currents, particularly the high-frequency currents. He put the master's touch to the research in this field by developing the principle of resonance in individual circuits through adjustment of capacity and inductance; the amplification of effects by inductive coupling of two tuned circuits, and the peculiar manifestations of resonance in a circuit tuned to a quarter of the wavelength of the energizing current. This latter development was a stroke of pure genius.

In the vibrating string, two loops measure a complete wavelength and one loop measures half a wavelength, since one of the loops is up when the other is down. Between the two loops is a nodal point which does not move. From the nodal point to the top of a loop is a quarter wavelength. Taking the quarter wavelength as a unit, one end is motionless and the other end swings through the greatest amplitude of vibration.

By tuning his coils to quarter wavelengths, one end of the coil, Tesla found, would be entirely inactive while the other end would swing through tremendous electrical activity. Here was a unique situation, one end of a small coil inert and the other end spouting a flood of sparks of hundreds of thousands or even millions of volts. In a physical analogy it seemed like the Niagara River reaching the edge of the precipice—and then its waters shooting mountain high in a gigantic fountain instead of falling into the chasm.

The quarter-wavelength coil is the electrical counterpart of the vibrating tine of the tuning fork, the ordinary clock pendulum, or the vibrating reed. Once accomplished, it was a simple thing—but its discovery was a work of genius. It was a development that could have come with certainty to a master mind working on broad principles, as Tesla was doing all his life, and only by the most improbable chance to those who without illumination were tinkering with gadgets and hoping to stumble on something out of which they could make a fortune.

A high-voltage coil with one dead end greatly simplified many problems. One of Tesla's big problems had been the finding of means to insulate the high-voltage secondary coil of transformers from the low-voltage primary which energized it. Tesla's discovery eliminated the voltage entirely from one end of the secondary so it could be connected directly to the primary or to the ground, while the other end continued to spout its lightning. It was for utilizing this situation that he developed the conical and pancake-shaped coils.

Tesla's laboratory was filled with a variety of coils. He discovered early in his researches that while operating a coil of a given wavelength, other coils in the laboratory, tuned either to this wavelength or one of its harmonics, would respond sympathetically by spouting a crown of sparks although not connected in any way to the operating coil.

Here was an example of transmission of energy to a distance through space. It was not necessary for Tesla to make a series of experiments to understand the implications of this situation. He was never lost in a new territory which he opened. His mind rose to such heights of understanding that he could survey a revealed world in a glance.

Tesla planned a spectacular demonstration of the new principle. He had his workmen string a wire on insulating supports on all four walls near the ceiling of the largest room in his laboratory. The wire was connected to one of his oscillators.

It was late at night when the installation was ready for the experiment. In order to make the test, Tesla prepared two tubes of glass about three feet long and a half-inch in diameter. He sealed one end of each, slightly evacuated the air from the tubes, and then sealed the other ends.

Tesla told the workmen he wanted the room completely darkened for the test, all lights out; and when he gave the signal he wanted the switch of his oscillator closed. "If my theory is correct," he explained, "when you close the switch these tubes will become swords of fire."

Walking to the middle of the room Tesla gave orders to turn out all lights. The laboratory was in pitch darkness. A workman stood with his hand on the switch of the oscillator.

"Now!" shouted Tesla.

Instantly the great room was flooded with brilliant but weird blue-white light and the workmen beheld the tall, slim figure of Tesla in the middle of the room waving vigorously what looked like two flaming swords. The two glass tubes glowed with an unearthly radiance, and he would parry and thrust with them as if he were in a double fencing match.

To the workmen in the laboratory, it was a common experience for Tesla to perform spectacular feats; but this went beyond all limits. He had previously lighted his electric vacuum lamps but they were always connected to coils that supplied them with electricity. Now they lighted without being connected to any source of electricity.

This demonstration, made in 1890, led to Tesla's adopting the technique as the permanent method of lighting his laboratories. The loop around the ceiling was always energized; and if anyone wished a light at any position, it was only necessary to take a glass tube and place it in any convenient location.

WHEN Tesla undertook the development of a new kind of electric light, he went to the sun for his model. He saw in the photosphere, or outer gaseous layer of the sun, light being created by the vibration of molecules. That was the theory then prevalent; and he sought to use the same method.

In the tremendous burst of revelation which he received in the park at Budapest as he gazed into the flaming orb of the setting sun, there had flashed into his mind, as we have seen, not only the marvelous invention of the rotary magnetic field and the many uses of multiple alternating currents, but also the grand generalization that everything in Nature operated on the principle of vibrations that corresponded to alternating currents. The host of inventions and discoveries which he made in all succeeding years had their roots, too, in that sublime experience.

In the sun, it was believed, light was created when the molecules were vibrated by heat. Tesla sought to improve on this method by vibrating the molecules by electrical forces. The sparks and electrical flames created by his high-voltage coils were associated, he believed, with molecular vibrations in the air. If he could bottle the gases of the air and set them in vibration electrically, they should produce light without heat, since the energy was supplied by cold electric currents.

Sir William Crookes, who, long before Edison, produced an incandescent electric light by sealing an electrically heated wire in a vacuum tube, had carried out an extended series of experiments in conducting electricity through the gases in glass vessels under a variety of conditions ranging from atmospheric pressure to the highest vacuum obtainable, and had produced some strange effects. Crookes used the high-voltage current produced by the old-fashioned induction coil.

Tesla expected that when he bottled the strange effects he had observed with his currents of extremely high frequency, he would produce manifestations radically different from those found by Crookes, or Geissler, who also worked in this field. In this he was not disappointed.

Four types of an entirely new kind of electric light were produced by Tesla, using electrically activated molecules of gas: 1. Tubes in which a solid body was rendered incandescent; 2. Tubes in which phosphorescent and fluorescent materials were caused to luminesce; 3. Tubes in which rarefied gases became luminous, and 4. Tubes in which luminosity was produced in gases at ordinary pressures.

Like Crookes, Tesla passed his high-frequency currents through gases at all pressures, from lowest-pressure vacuum to normal atmospheric pressure, and obtained brilliant luminous effects exceeding anything previously attained. He substituted for air in his tubes other gases, including mercury vapor, and observed the peculiar color and other effects they yielded.

Noting the variety of colors the various gases, and even air, showed under different pressures, Tesla suspected that not all of the energy radiated was given off as visible light, but that some of it emanated as black light. Testing this hypothesis, he placed sulphide of zinc and other phosphorescent and fluorescent materials in his tubes and caused them to glow. In these experiments (they were made in 1889) Tesla laid the foundation for our most recently developed type of highly efficient lamps used in fluorescent lighting which are generally believed to have been invented in recent years. This system of utilizing the wasted ultra-violet or invisible black light by changing it to visible light by means of phosphorescent substances is Tesla's invention. Roentgen was using similar tubes, but of plain glass and the fluorescent substance on a table in his laboratory when, a half-dozen years later, he discovered X-rays. Tesla invented, also, the neon-tube type of lamp, and even bent his tubes to form letters and geometrical shapes, as is done in neon-tube signs. This is true in spite of some antecedent and concurrent laboratory experiments by Crookes and J. J. Thompson, neither of whom developed any lamps or practical applications.

Tesla had discovered early in 1890 that his high-frequency currents had properties so different from the ordinary induction-coil, or spark-coil, currents, that he was able to light his tubes just as well, and sometimes even better, with only one wire connecting them with the high-tension transformer, the return circuit being effected wirelessly through space.

In working with types of lamps consisting of tubes in the center of which there was a conducting wire, and with the tube filled with air under a partial vacuum, Tesla discovered that the gas would serve as a better conductor of the high-frequency current than the wire. From this observation he was able to develop many spectacular experiments which appeared to violate the most fundamental laws of electricity. He was able to short circuit lamps and other apparatus with heavy bars of metal which, with ordinary currents, would completely deprive the devices of electricity so they would be unable to operate. However, with his high-frequency currents, the lamps would light and the devices operate just as if the short-circuiting bar were not present.

One of his startling experiments consisted of placing a long glass tube partially evacuated of its air inside a slightly longer copper tube with a closed end. A slit was cut in the copper tube in its central section so the tube inside would be seen. When the copper tube was connected in the high-frequency circuit, the air in the tube was brilliantly illuminated; but no evidence could be found of any current flowing through the short-circuiting copper shell. The electricity preferred to pass through the glass tube, by induction, to the enclosed partially evacuated air, pass through the low-pressure air for the full length of the tube, and then pass out the other end by induction, rather than traverse the complete metal path in the surrounding metal tube.

We have then, [said Tesla], as far as we can now see, in the gas a conductor which is capable of transmitting electric impulses of any frequency which we may be able to produce. Could the frequency be brought high enough, then a queer system of distribution, which would be likely to interest gas companies, might be realized; metal pipes filled with gas—the metal being the insulator and the gas the conductor—supplying phosphorescent bulbs, or perhaps devices not yet invented.

This remarkable conductivity of gases, including the air, at low pressures, later led Tesla to suggest, in a published statement in 1914, a system of lighting on a terrestrial scale in which he proposed to treat the whole earth, with its surrounding atmosphere, as if it were a single lamp.

The atmosphere is under the greatest pressure at the surface of the earth, owing to the weight of the overlying air. As we go higher in the air there are increasing amounts below us and less above, so, the greater the elevation, the lower is the pressure of the air.

At higher altitudes the gases in the atmosphere are in the same condition as the air in the partially evacuated tubes he prepared in his laboratory, Tesla explained, and therefore it would serve as an excellent conductor of high-frequency currents. The aurora borealis is a natural example of the effect Tesla sought, and it is produced by Nature as Tesla planned; but this was not known when he evolved his idea.

The flow of a sufficient amount of the electricity in the right form through the upper regions of the atmosphere would cause the air to become luminous. The whole earth would be transformed into a giant lamp, with the night sky completely illuminated. It would be unnecessary, he pointed out, to use any lamps along streets, roads or other outdoor areas, except during periods in which storms or low clouds prevailed. Ocean travel would be made safer and more pleasant, for the sky over the whole ocean would be illuminated, making the night as bright as day.

The methods by which Tesla intended to conduct his high-frequency currents to the upper air have not been published. When he outlined the project, he stated that the plan did not present any difficulties that could not be handled in a practical way. This meant that he had definite means for accomplishing his purpose.

The air, he stated, possesses a high degree of conductivity for high-frequency currents at an altitude of 35,000 feet, but could be used effectively at lower atitudes. The accuracy of Tesla's prediction with respect to the conductivity of the upper air is attested by one of the problems encountered today in the operation of airplanes at altitudes even lower than 25,000 feet. The ignition system, carrying high-voltage currents to the spark plugs in the airplane engines, which explodes the gas in the cylinders, has been giving trouble at the higher altitudes because the electricity escapes with a great deal of freedom into the surrounding air. At lower altitudes the air is an excellent insulator, especially for direct current and low-frequency currents, but, as Tesla discovered, at the higher altitudes where low pressures prevail it becomes an excellent conductor for the high-frequency currents. The wires leading to spark plugs become surrounded by a corona, or electrical halo, which indicates the escape of the current. This interferes with the efficiency, if it does not entirely prevent the operation, of devices employing high-frequency or high-potential currents, such as radio apparatus.*

This proposal by Tesla to transform the earth into a giant lamp was again referred to by him in the twenties. At this time he was without funds for carrying on experimental work, and, as he never announced details until after he had tested them in practice, he withheld a disclosure of his methods. He was hopeful, however, that he would soon secure money enough to permit him to test his plan.

The author bombarded Tesla with questions in an effort to learn the general plan he had in mind. Tesla was adamant.

"If I should answer three more of your questions you would know as much about my plan as I do," he replied.

"Nevertheless, Dr. Tesla," I replied, "I am going to outline in my article the only plan that appears to me to be feasible under our known physical laws, and you can deny or affirm it. Your molecular bombardment tubes are prolific producers of ultra-violet and X-rays and could produce a powerful beam of this radiation which would ionize the air through great distances. When these rays pass through the air they ionize it, making it a good conductor of electricity of all kinds at sufficiently high voltages. By producing such a beam on a high mountain and directing it upward this would provide a conducting path through the air to any height desired. You could then send your high-frequency currents to the upper air without leaving the ground."

"If you publish that," said Tesla, "it must appear as your plan, not mine."

The article was published with the foregoing speculation in

* Since Tesla discovered that metal wires and rods which act as excellent conductors for direct and low-frequency currents can act as excellent insulators for his high-frequency currents, it is obvious that the common suggestion made for delivering a current to the upper air by means of metal cables suspended from balloons is entirely impractical.

it; but neither affirmation nor denial was forthcoming from the inventor, and nothing more can be said in its favor. Tesla may have had a simpler and more practical plan in mind.*

There was one other plan which Tesla discussed on a number of occasions when considering terrestrial electrical conditions, and which he may have had in mind in this connection. He pointed out that the earth is a good conductor of electricity and the upper air is also a good conductor, while the intervening lower stratum of air is an insulator for many kinds of current. This combination provides what is known as a condenser, a device which will store and discharge electricity. By charging the earth, the upper air would become charged by induction. When our spinning earth was so transformed into a terrestrial Leyden jar, it could be alternately charged and discharged, so that a current would flow both in the upper air and in the ground, producing the electrical flow which would cause the upper air to become self-luminous. Tesla, however, never became quite so specific in applying the condenser plan to this problem as the preceding sentence indicates. His plan may still exist in his papers, which, at the present writing, are sealed against inspection except by Government officials.

OUT of the almost empty space in a six-inch vacuum tube Tesla succeeded in extracting at least five epoch-making discoveries. Tesla's lamp was more prolific in producing wonders than the Aladdin's lamp of the *Arabian Nights*. He gave his "magic" lamp to science fifty years ago. This magic talisman was Tesla's carbon-button lamp which, apart from the other discoveries that came of it, was in itself, just as a lamp, a brilliant scientific discovery—and still remains unused. Edison developed the practical incandescent filament electric lamp and was entitled to, and receives, a tremendous amount of credit for his accomplishment.

---

* Since completing this volume the author has learned that Tesla planned to install a bank of powerful ultra-violet lamps on top of his tower at Wardencliff (cf. p. 205), and had the upper platform designed to receive them.

Tesla invented an absolutely original type of lamp, the incandescent-button lamp, which gives twenty times as much light for the same amount of current consumed; and his contribution remains practically unknown.

The carbon-button type of lamp was described by Tesla in his lecture before the American Institute of Engineers in New York in May, 1891, and further developments were presented in the lectures which he gave in England and France in February and March, 1892. In his New York lecture he said:

Electrostatic effects are in many ways available for the production of light. For instance, we may place a body of some refractory material in a closed, and preferably in a more or less air exhausted, globe, connect it to a source of high, rapidly alternating potential, causing the molecules of the gas to strike it many times a second at enormous speeds, and in this way, with trillions of invisible hammers, pound it until it gets incandescent. Or we may place a body in a very highly exhausted globe, and by employing very high frequencies and potentials maintain it at any desired degree of incandescence.

He made a vast number of experiments with this carbon-button lamp and gave a description of the most significant ones in his lecture before the English and French scientific societies in the spring of 1892. It was, however, only one of the many new types of lamps and other important developments which he included in this spectacular presentation of his work.

The carbon-button lamps were of very simple construction. Basically they consisted of a spherical glass globe three to six inches in diameter, in the center of which was a piece of solid refractory material mounted on the end of a wire which protruded through the globe and served as a single-wire connection with the source of high-frequency currents. The globe contained rarefied air.

When the high-frequency current was connected with the lamp, molecules of the air in the globe, coming in contact with the central button, became charged and were repelled at high velocity to the glass globe where they lost their charge and were then repelled back at equally high velocity, striking the button. Millions of millions of such processes each second caused the button to become heated to incandescence.

In these simple glass globes Tesla was able to produce extremely high temperatures, the upper limit of which seemed to be determined by the amount of current used. He was able to vaporize carbon directly into a gas, observing that the liquid state was so unstable it could not exist. Zirconia, the most heat-resistant substance known, could be melted instantly. He tried diamonds and rubies as buttons—and they too were vaporized. When using the device as a lamp it was not his desire to melt the substances; but he always carried experiments to their upper and lower limits. Carborundum, he observed, was so refractory that it was possible when using buttons made of this material (calcium carbide) to run the lamps at higher current densities than was possible with other substances. Carborundum did not vaporize so readily, nor did it make deposits on the inside of the globe.

Tesla thus evolved a technique in operating the lamps whereby the incandescent button transferred its heat energy to the molecules of the very small amount of gas in the tube so that they became a source of light, thus causing the lamps to function like the sun, the button being the massive body of the sun and the surrounding gas like the photosphere, or atmospheric light-emitting layer, of that body.

Tesla had a keen sense of dramatic values, but quite apart from this he undoubtedly enjoyed a unique satisfaction when he was able to light this miniature sun in the currents that he passed through his body—high-frequency currents of hundreds of thousands of volts. With one hand grasping a terminal of his high-frequency transformer and the other holding aloft this bulb containing an incandescent miniature sun which he had created —posing like the Statue of Liberty—he was able to make his new lamp radiate its brilliant illumination. Here, you might say, was the superman manifesting his ultramundane accomplishments. In addition, there was a satisfaction which was associated purely with the plane of ordinary mortals. Edison had laughed at his plan for developing the alternating-current system, and had declared that these currents were not only useless but deadly. Surely, this was an adequate answer; Tesla would let Nature make his replies.

Observing this working model of the incandescent sun which he could hold in his hand, Tesla was quick to see many of the implications of its phenomena. Every electrical wave that surged through the tiny central bead caused a shower of particles to radiate from it at tremendous velocity and strike the surrounding glass globe, only to be reflected back to the bead. The sun, Tesla reasoned, is an incandescent body that carries a high electrical charge and it, too, will emit vast showers of tiny particles, each carrying great energy because of its extremely high velocity. In the case of the sun, and other stars like it, there was no glass globe to act as a barrier, so the showers of particles continued out into the vast realms of surrounding space.

All space was filled with these particles and they were continually bombarding the earth, blasting matter wherever they struck, just as they did in his globes. He had seen this process take place in his globes, where the most refractory carbon beads could be shattered into atomic dust by the bombardment of the electrified particles.

He sought to detect these particles striking the earth: one of the manifestations of this bombardment, he declared, was the aurora borealis. The records of the experimental methods by which he detected these rays are not available; but he published an announcement that he had detected them, measured their energy, and found that they moved with tremendously high velocities imparted to them by the hundreds of millions of volts potential of the sun.

Neither the scientists nor the general public in the early nineties were in a mood for such fantastic figures, or for any claim that the earth was bombarded by such destructive rays. It would be describing the situation in very conservative fashion to state that Tesla's report was not taken seriously.

When, however, the French physicist, Henri Becquerel, in 1896, discovered the mysterious rays emitted by uranium, and subsequent investigations, culminating with the discovery by Pierre and Marie Curie, in Paris, of radium, whose atoms were exploding spontaneously without apparent cause, Tesla was able to point to his cosmic rays as the simple cause of the radioactivity of radium, thorium, uranium and other substances. And he predicted that other substances would also be found to be made radioactive by bombardment with these rays. The victory for Tesla, however, was only temporary, for the scientific world did not accept his theory. Nevertheless, Tesla was a better prophet than he knew, or anyone else suspected.

Thirty years later Dr. Robert A. Millikan rediscovered these rays, believing them to be vibratory in character like light, and was followed by Dr. Arthur H. Compton, who proved the existence of cosmic rays consisting of high-velocity particles of matter, just as Tesla described them. They started by finding energies of ten million volts; and today the energies are far up in the billions and even trillions of electron volts. And these and other investigators describe these rays as shattering atoms of matter producing showers of débris—just as Tesla predicted.

In 1934, Frederick Joliot, son-in-law of the Curies, discovered that artificial radioactivity was produced in ordinary materials by bombarding them with particles in just the manner which Tesla described. Joliot received the Nobel Prize for his discovery; no one gives any credit to Tesla.

Tesla's molecular-bombardment lamp was the ancestor of another very modern development—the atom-smashing cyclotron. The cyclotron, developed by E. O. Lawrence, of the University of California, during the past twenty years, is a device in which electrified particles are whirled in a magnetic field in a circular chamber until they reach a very high velocity, and are then led out of the chamber in a narrow stream. The giant machine, with a magnet as high as a house, partially completed at present writing, will emit so powerful a beam of charged particles that, according to Prof. Lawrence, if allowed to impinge on a building brick they will totally disintegrate it. The smaller models were used to bombard a variety of substances to render them radioactive, to disintegrate them or transmute their atoms into those of other elements.

The small glass globe, six inches or less in diameter, holding Tesla's molecular-bombardment lamp produced exactly this same disintegrating effect on solid matter, probably with a more intensified effect than any atom-smashing cyclotron now in existence—despite their tremendous size. (Even small ones weigh twenty tons.)

In describing one of the experiments with his lamp, one in which a ruby was mounted in a carbon button, Tesla said:

It was found, among other things, that in such cases, no matter where the bombardment began, just as soon as a high temperature was reached there was generally one of the bodies which seemed to take most of the bombardment upon itself, the other, or others, being thereby relieved. This quality appeared to depend principally on the point of fusion, and on the facility with which the body was "evaporated," or, generally speaking, disintegrated—meaning by the latter term not only the throwing off of atoms, but likewise of larger lumps. The observation made was in accordance with generally accepted notions. In a highly exhausted bulb electricity is carried off from the electrode by independent carriers, which are partly atoms, or molecules, of the residual atmosphere, and partly the atoms, molecules, or lumps thrown off from the electrode. If the electrode is composed of bodies of different character, and if one of these is more easily disintegrated than the others, most of the electricity supplied is carried off from that body, which is then brought to a higher temperature than the others, and this the more, as upon an increase of the temperature the body is still more easily disintegrated.

151

Substances which resisted melting in temperatures attainable in laboratory furnaces of that day were easily disintegrated in Tesla's simple-lamp disintegrator, which provided a powerful beam of disintegrating particles by having them concentrated from all directions by a spherical reflector (the globe of his lamp), a kind a three-dimension burning glass, but operating with electrified particles instead of heat rays. It accomplished the same effect as the heavy atom disintegrators of today, but much more efficiently in a globe so light in weight it almost floated off in air. Its simplicity and efficiency is further increased by the fact that it causes the substance that is being disintegrated to supply the particles by which the disintegration is effected.

There is one more very modern discovery of great importance embodied in Tesla's molecular-bombardment lamp—the point electron miscroscope, which provides magnifications of a million diameters, or ten to twenty times more powerful than the better known electron microscope—which in turn is capable of magnifications up to fifty times greater than the optical microscope.

In the point electron microscope, electrified particles shoot out in straight lines from a tiny active spot on a piece of substance kept at a high potential, and reproduce on the spherical surface of a glass globe the pattern of the microscopically small area from which the particles are issuing. The size of the glass sphere furnishes the only limit to the degree of magnification that can be obtained; the greater the radius, the greater the magnification. Since electrons are smaller than light waves, objects too small to be seen by light waves can be tremendously enlarged by the patterns produced by the emitted electrons.

Tesla produced on the surface of the spherical globe of his lamp phosphorescent images of what was taking place on the disintegrating button when he used extremely high vacuum. He described this effect in his lectures in the spring of 1892, and his description will stand with hardly a change in a word for a description of the million-magnification point electron microscope. Quoting from his lecture:

To the eye the electrode appears uniformly brilliant, but there are upon it points constantly shifting and wandering about, of a temperature far above the mean, and this materially hastens the process of deterioration. . . . Exhaust a bulb to a very high degree, so that with a fairly high potential the discharge cannot pass—that is, not a luminous one, for a weak invisible discharge occurs always, in all probability. Now raise slowly and carefully the potential, leaving the primary current no more than for an instant. At a certain point, two, three or half a dozen phosphorescent spots will appear on the globe. These places of the glass are evidently more violently bombarded than the others, this being due to the unevenly distributed electric density, necessitated, of course, by sharp projections, or, generally speaking, irregularities of the electrode. But the luminous patches are constantly changing in position, which is especially well observed if one manages to produce very few, and this indicates that the configuration of the electrode is rapidly changing.

It would be an act of simple justice if in the future scientists would extend credit to Tesla for being the one who discovered the electron miscroscope. There is no reduction in the glory due him because he did not specifically describe the electron, then unknown, in its operations, but assumed the effect was due to electrically charged atoms.

When Tesla studied the performance of various models of this and his other gaseous lamps, he observed that the output of visible light changed under various operating conditions. He knew they gave off both visible and invisible rays. He used a variety of phosphorescent and fluorescent substances for detecting the ultra-violet or black light. Usually, the changes in the visible and ultra-violet light about balanced each other; as one increased the other decreased, with the remainder of the energy accounted for by heat losses. In his molecular-bombardment lamp he found, he reported in his 1892 lectures, "visible light, black light and a very special radiation." He was experimenting with this radiation which, he reported, produced shadowgraph pictures on plates in metal containers, in his laboratory when it was destroyed by fire in March, 1895.

This "very special radiation" was not further described at that time in published articles; but when Prof. Wilhelm Konrad Roentgen, in Germany, in December, 1895, announced the discovery of X-rays, Tesla was able immediately to reproduce the same results by means of his "very special radiation," indicating that these and X-rays had very similar properties although produced in somewhat different ways. Immediately upon reading Roentgen's announcement, Tesla forwarded to the German scientist shadowgraph pictures produced by his "very special radiation." Roentgen replied: "The pictures are very interesting. If you would only be so kind as to disclose the manner in which you obtained them."

Tesla did not consider that this situation gave him any priority in the discovery of X-rays, nor did he ever advance any claims; but he immediately started an extensive series of investigations into their nature. While others were trying to coax out of the type of tube used by Roentgen enough X-rays to take shadow photographs through such thin structures as the hands and feet held very close to the bulb, Tesla was taking photographs through the skull at a distance of forty feet from the tube. He elsewhere described at this time an unidentified type of radiation coming from a spark gap, when a heavy current was passed, that was not a transverse wave like light, or Hertzian waves, and could not be stopped by interposing metal plates.

Tesla, thus, in one lecture reporting his investigations covering a period of two years, offered to the world—in addition to his new electric vacuum lamps, his highly efficient incandescent lamp, and his high-frequency and high-potential currents and apparatus—at least five outstanding scientific discoveries: 1. Cosmic rays; 2. Artificial radioactivity; 3. Disintegrating beam of electrified particles, or atom smasher; 4. Electron microscope; and 5. "Very special radiation" (X-rays).

At least four of these innovations, when "rediscovered" up to forty years later, won Nobel Prizes for others; and Tesla's name is never mentioned in connection with them.

Yet Tesla's lifetime work was hardly well started!

# *TEN*

---

TESLA had a remarkable ability for carrying on simultaneously a number of widely different lines of scientific research. While pursuing his studies of high-frequency electrical oscillations with all of their ramifications from vacuum lamps to radio, he was also investigating mechanical vibrations; and he had a rare foresight into the many useful applications to which they could be put, and which have since been realized.

Tesla never did things by halves. Almost everything he attempted went off like a flash of lightning with a very satisfactory resounding clap of thunder following. Even when he did not so plan events, they appeared to fashion themselves into spectacular climaxes. In 1896 while his fame was still on the ascendant he planned a nice quiet little vibration experiment in his Houston Street laboratory. Since he had moved into these quarters in 1895, the place had established a reputation for itself because of the peculiar noises and lights that emanated from it at all hours of the day and night, and because it was constantly being visited by the most famous people in the country.

The quiet little vibration experiment produced an earthquake, a real earthquake in which people and buildings and everything in them got a more tremendous shaking than they did in any of the natural earthquakes that have visited the metropolis. In an area of a dozen square city blocks, occupied by hundreds of buildings housing tens of thousands of persons, there was a sudden roaring and shaking, shattering of panes of glass, breaking of steam, gas and water pipes. Pandemonium reigned as small objects danced around rooms, plaster descended from walls and ceilings, and pieces of machinery weighing tons were moved from their bolted anchorages and shifted to awkward spots in factory lofts.

"It was all caused, quite unexpectedly, by a little piece of apparatus you could slip in your pocket," said Tesla.

The device that precipitated the sudden crisis had been used for a long time by Tesla as a toy to amuse his friends. It was a mechanical oscillator, and was used to produce vibrations. The motor-driven device that the barber straps on his hand to give a patron an "electric massage" is a descendant of Tesla's mechanical oscillator. There is, of course, nothing electric about an "electric massage" except the power used to produce vibrations which are transmitted through the barber's fingers to the scalp.

Tesla developed in the early nineties a mechanical-electrical oscillator for the generation of high-frequency alternating currents. The driving engine produced on a shaft simple reciprocating motion that was not changed to rotary motion. Mounted on either end of the shaft was a coil of many turns of wire that moved back and forth with high frequency between the poles of electromagnets, and in this way generated high-frequency alternating currents.

The engine was claimed by Tesla to have a very high efficiency compared to the common type of engine, which changed reciprocating to rotary motion by means of a crank shaft. It had no valves or other moving parts, except the reciprocating piston with its attached shaft and coils, so that mechanical losses were very low. It maintained such an extremely high order of constancy of speed, he stated, that the alternating current generated by the oscillator could be used to drive clocks, without any pendulum or balance-wheel control mechanisms, and they would keep time more accurately than the sun.

This engine may have had industrial possibilities but Tesla was not interested in them. To him it was just a convenient way of producing a high-frequency alternating current constant in frequency and voltage, or mechanical vibrations, if used without the electrical parts. He operated the engine on compressed air and also by steam at 320 pounds and also at 80 pounds pressure.

While perfecting this device, he had opportunity to observe interesting effects produced by vibration. These were objectionable in the engine when it was used as a dynamo, so he adopted suitable measures to eliminate or suppress them. The vibrations as such, however, interested him. Although they were detrimental to the machine, he found their physiological effects were, at times, quite pleasant. Later he built a small mechanical oscillator driven by compressed air which was designed for no other purpose than to produce vibrations. He built a platform insulated from the floor by rubber and cork. He then mounted the oscillator on the under side of the platform. The purpose of the rubber and cork under the platform was to keep the vibrations from leaking into the building and thereby reducing the effect on the platform. Visitors found this vibrating platform one of the most interesting of the great array of fascinating and fantastic exhibits with which he dazzled the society folk who flocked to his laboratory.

Great hopes were entertained by Tesla of applying these vibrations for therapeutic and health-improving effects. He had opportunity to observe, through his own experience and that of his employees, that they produce some very definite physiological actions.

Samuel Clemens, better known to the public as "Mark Twain," and Tesla were close friends. Clemens was a frequent visitor to the Tesla laboratory. Tesla had been playing with his vibratory mechanism for some time, and had learned a good deal about the results that followed from varying doses of vibration, when one evening Clemens dropped in.

Clemens, on learning about the new mechanism, wanted to experience its vitalizing vibrations. He stood on the platform while the oscillator set it into operation. He was thrilled by the new experience. He was full of adjectives. "This gives you vigor and vitality," he exclaimed. After he had been on the platform for a while Tesla advised him: "You have had enough, Mr. Clemens. You had better come down now."

"Not by a jugful," replied Clemens. "I am enjoying myself."

"But you had better come down, Mr. Clemens. It is best that you do so," insisted Tesla.

"You couldn't get me off this with a derrick," laughed Clemens.

"Remember, I am advising you, Mr. Clemens."

"I'm having the time of my life. I'm going to stay right up here and enjoy myself. Look here, Tesla, you don't appreciate what a wonderful device you have here to give a lift to tired humanity. . . ." Clemens continued along this line for several minutes. Suddenly he stopped talking, bit his lower lip, straightened his body and stalked stiffly but suddenly from the platform.

"Quick, Tesla! Where is it?" snapped Clemens, half begging, half demanding.

"Right over here, through that little door in the corner," said Tesla. "And remember, Mr. Clemens, I advised you to come down some time ago," he called after the rapidly moving figure.

The laxative effect of the vibrator was an old story to the members of the laboratory staff.

Tesla pursued his studies of mechanical vibrations in many directions. This was almost a virgin field for scientific research. Scarcely any fundamental research had been done in the field since Pythagoras, twenty-five hundred years before, had established the science of music through his study of vibrating strings; and many of the wonders with which Tesla had startled the world in the field of high-frequency and high-potential currents had grown out of his simple secret for tuning electrical circuits so that the electricity vibrated in resonance with its circuit. He now visualized mechanical vibrations building up resonance conditions in the same way, to produce effects of tremendous magnitude on physical objects.

In order to carry out what he expected to be some minor and very small-scale experiments, he screwed the base of one of his small mechanical oscillators to an iron supporting pillar in the middle of his laboratory and set it into oscillation. It had been his observation that it took some time to build up its maximum speed of vibration. The longer it operated the faster the tempo it attained. He had noticed that all objects did not respond in the same way to vibrations. One of the many objects around the laboratory would suddenly go into violent vibration as it came into resonance with the fundamental vibration of the oscillator or some harmonic of it. As the period of the oscillator changed, the first object would stop and some other object in resonance with the new rate would start vibrating. The reason for this selective response was very clear to Tesla, but he had never previously had the opportunity to observe the phenomenon on a really large scale.

Tesla's laboratory was on an upper floor of a loft building. It was on the north side of Houston Street, and the second house east of Mulberry Street. About three hundred feet south of Houston Street on the east side of Mulberry Street was the long, four-story red-brick building famous as Police Headquarters. Throughout the neighborhood there were many loft buildings ranging from five to ten stories in height, occupied by factories of all kinds. Sandwiched between them were the small narrow tenement houses of a densely packed Italian population. A few blocks to the south was Chinatown, a few blocks to the west was the garment-trades area, a short distance to the east was a densely crowded tenement-house district.

It was in this highly variegated neighborhood that Tesla unexpectedly staged a spectacular demonstration of the properties of sustained powerful vibrations. The surrounding population knew about Tesla's laboratory, knew that it was a place where strange, magical, mysterious events took place and where an equally strange man was doing fearful and wonderful things with that tremendously dangerous secret agent known as electricity. Tesla, they knew, was a man who was to be both venerated and feared, and they did a much better job of fearing than of venerating him.

Quite unmindful of what anyone thought about him, Tesla

carried on his vibration and all other experiments. Just what experiment he had in mind on this particular morning will never be known. He busied himself with preparations for it while his oscillator on the supporting iron pillar of the structure kept building up an ever higher frequency of vibrations. He noted that every now and then some heavy piece of apparatus would vibrate sharply, the floor under him would rumble for a second or two—that a window pane would sing audibly, and other similar transient events would happen—all of which was quite familiar to him. These observations told him that his oscillator was tuning up nicely, and he probably wondered why he had not tried it firmly attached to a solid building support before.

Things were not going so well in the neighborhood, however. Down in Police Headquarters in Mulberry Street the "cops" were quite familiar with strange sounds and lights coming from the Tesla laboratory. They could hear clearly the sharp snapping of the lightnings created by his coils. If anything queer was happening in the neighborhood, they knew that Tesla was in back of it in some way or other.

On this particular morning the cops were surprised to feel the building rumbling under their feet. Chairs moved across floors with no one near them. Objects on the officers' desks danced about and the desks themselves moved. It must be an earthquake! It grew stronger. Chunks of plaster fell from the ceilings. A flood of water ran down one of the stairs from a broken pipe. The windows started to vibrate with a shrill note that grew more intense. Some of the windows shattered.

"That isn't an earthquake," shouted one of the officers, "it's that blankety-blank Tesla. Get up there quickly," he called to a squad of men, "and stop him. Use force if you have to, but stop him. He'll wreck the city."

The officers started on a run for the building around the corner. Pouring into the streets were many scores of people excitedly leaving near-by tenement and factory buildings, believing an earthquake had caused the smashing of windows, breaking of pipes, moving of furniture and the strange vibrations.

Without waiting for the slow-pokey elevator, the cops rushed up the stairs—and as they did so they felt the building vibrate even more strongly than did police headquarters. There was a sense of impending doom—that the whole building would disintegrate—and their fears were not relieved by the sound of smashing glass and the queer roars and screams that came from the walls and floors.

Could they reach Tesla's laboratory in time to stop him? Or would the building tumble down on their heads and everyone in it be buried in the ruins, and probably every building in the neighborhood? Maybe he was making the whole earth shake in this way! Would this madman be destroying the world? It was destroyed once before by water. Maybe this time it would be destroyed by that agent of the devil that they call electricity!

Just as the cops rushed into Tesla's laboratory to tackle—they knew not what—the vibrations stopped and they beheld a strange sight. They arrived just in time to see the tall gaunt figure of the inventor swing a heavy sledge hammer and shatter a small iron contraption mounted on the post in the middle of the room. Pandemonium gave way to a deep, heavy silence.

Tesla was the first to break the silence. Resting his sledge hammer against the pillar, he turned his tall, lean, coatless figure to the cops. He was always self-possessed, always a commanding presence—an effect that could in no way be attributed to his slender build, but seemed more to emanate from his eyes. Bowing from the waist in his courtly manner, he addressed the policemen, who were too out of breath to speak, and probably overawed into silence by their fantastic experience.

"Gentlemen," he said, "I am sorry, but you are just a trifle too late to witness my experiment. I found it necessary to stop it suddenly and unexpectedly and in an unusual way just as you entered. If you will come around this evening I will have another oscillator attached to this platform and each of you can stand on it. You will, I am sure, find it a most interesting and pleasurable experience. Now you must leave, for I have many things to do. Good day, gentlemen."

George Scherff, Tesla's secretary, was standing near by when Tesla so dramatically smashed his earthquake maker. Tesla never told the story beyond this point, and Mr. Scherff declares he does not recall what the response of the cops was. Imagination must furnish the finale to the story.

At the moment, though, Tesla was quite sincere in his attitude. He had no idea of what had happened elsewhere in the neighborhood as a result of his experiment, but the effect on his own laboratory had been sufficiently threatening to cause him to halt it suddenly. When he learned the details, however, he was convinced that he was correct in his belief that the field of mechanical vibrations was rich with opportunities for scientific investigation. We have no records available of any further major experiments with vibration in that laboratory. Perhaps the Police and Building Departments had offered some emphatic suggestions to him concerning experiments of this nature.

Tesla's observations in this experiment were limited to what took place on the floor of the building in which his laboratory was located, but apparently very little happened there until a great deal had happened elsewhere. The oscillator was firmly fixed to a supporting column and there were similar supporting columns directly under it on each floor down to the foundations. The vibrations were transmitted through the columns to the ground. This section of the city is built on deep sand that extends down some hundreds of feet before bed rock is reached. It is well known to seismologists that earthquake vibrations are transmitted by sand with much greater intensity than they are by rock. The ground under the building and around it was, therefore, an excellent transmitter of mechanical vibrations, which spread out in all directions. They may have reached a mile or more. They were more intense, of course, near their source and became weaker as the distance increased. However, even weak vibrations that are sustained can build up surprisingly large effects when they are absorbed by an object with which they are in resonance. A distant object in resonance can be thrown into strong vibration whereas a much nearer object not in resonance will be left unaffected.

It was this selective resonance that was, apparently, operating in Tesla's experiment. Buildings other than his own came into resonance with the increasing tempo of his oscillator long before his own building was affected. After the pandemonium was under way for some time elsewhere and the higher frequencies were reached, his immediate surroundings started to come into resonance.

When resonance is reached the effects follow instantly and powerfully. Tesla knew this, so when he observed dangerous resonance effects developing in his building he realized he had to act fast. The oscillator was being operated by compressed air supplied by a motor-driven compressor that fed the air into a tank, where it was stored under pressure. Even if the motor were shut off, there was plenty of air in the tank to keep the oscillator going for many minutes—and in that time the building could be completely wrecked and reduced to a pile of débris. With the vibrations reaching this dangerous amplitude, there was no time to try to disconnect the vibrator from the air line or to do anything about releasing the air from the tank. There was time for only one thing, and Tesla did that. He grabbed the near-by sledge hammer and took a mighty swing at the oscillator in hopes of putting it out of operation. He succeeded in his first attempt.

The device was made of cast iron and was of rugged construction. There were no delicate parts that could be easily damaged. Tesla has never published a description of the device, but its construction was principally that of a piston which moved back and forth inside a cast-iron cylinder. The only way to stop it from operating was to smash the outer cylinder. Fortunately, that is what happened from the first blow.

*163*

As Tesla turned around after delivering this lucky blow and beheld the visiting policemen, he could not understand the reason for their visit. The dangerous vibrations had developed in his building only within the preceding minute, and the policemen would not have had time to plan a visit in connection with them, he figured, so they must have come for some other less critical purpose, and therefore he proposed to dismiss them until a more opportune moment.

Tesla related this experience to me when I asked the inventor's opinion of a plan that I had suggested some time previously to Elmer Sperry, Jr., son of the famous inventor of many gyroscope devices. When a heavy gyroscope, such as is used in stabilizing ships, is forced to turn on its axis, it transmits a powerful downward thrust through the bearings in which the supporting gimbal is mounted. If a battery of such gyroscopes were mounted in regions where severe earthquakes take place it would transmit thrusts to the ground at equally timed intervals and build up resonance vibrations in the strata of the earth that would cause earthquake strains to be released while they were of small magnitude, thus producing very small earthquakes instead of letting the strains build up to large magnitudes which, when they let go, would cause devastating earthquakes.

The idea made a strong appeal to Tesla; and in his discussion, after telling me of the experience here related, he further declared that he had so far developed his study of vibrations that he could establish a new science of "telegeodynamics" which would deal not only with the transmission of powerful impulses through the earth to distant points to produce effects of large magnitude—in addition, he could use the same principles to detect distant objects. In the later thirties, before the outbreak of the war, he declared that he could apply these principles for the detection of submarines or other ships at a distance, even though they were lying at anchor and no engines operating on them.

His system of telegeodynamics, using mechanical vibrations, Tesla declared, would make it possible to determine the physical constant of the earth and to locate ore deposits far beneath the surface. This latter prediction has since been fulfilled, for many oil fields have been discovered by studying the vibrations reflected from sub-surface strata.

"So powerful are the effects of the telegeodynamic oscillator," said Tesla in reviewing the subject in the thirties, "that I could now go over to the Empire State Building and reduce it to a tangled mass of wreckage in a very short time. I could accomplish this result with utmost certainty and without any difficulty whatever. I would use a small mechanical vibrating device, an engine so small you could slip it in your pocket. I could attach it to any part of the building, start it in operation, allow it twelve to thirteen minutes to come to full resonance. The building would first respond with gentle tremors, and the vibrations would then become so powerful that the whole structure would go into resonant oscillations of such great amplitude and power that rivets in the steel beams would be loosened and sheared. The outer stone coating would be thrown off and then the skeleton steel structure would collapse in all its parts. It would take about 2.5 * horsepower to drive the oscillator to produce this effect."

Tesla developed his inventions to the point at which they were spectacular performers before they were demonstrated to the public. When presented, the performance always greatly exceeded the promise. This was the case with his first public demonstration of "wireless," but he complicated the situation by coupling with his radio invention another new idea—the robot.

Tesla staged his demonstration in the great auditorium of Madison Square Garden, then on the north side of Madison Square, in September, 1898, as part of the first annual Electrical Exhibition. He had a large tank built in the center of the arena

* This figure may have been .25 or 2.5 horsepower. The notes are old and somewhat indistinct. Memory favors the latter figure.

and in this he placed an iron-hulled boat a few feet long, shaped like an ark, which he operated by remote control by means of his wireless system.

Extending upward from the center of the roof of the boat was a slender metal rod a few feet high which served as an antenna, or aerial, for receiving the wireless wave. Near the bow and stern were two small metal tubes about a foot high surmounted by small electric lamps. The interior of the hull was packed with a radio receiving set and a variety of motor-driven mechanisms which put into effect the operating orders sent to the boat by wireless waves. There was a motor for propelling the boat and another motor for operating the servo-mechanism, or mechanical brain, that interpreted the orders coming from the wireless receiving set and translated them into mechanical motions, which included steering the boat in any direction, making it stop, start, go forward or backward, or light either lamp. The boat could thus be put through the most complicated maneuvers.

Anyone attending the exhibition could call the maneuver for the boat, and Tesla, with a few touches on a telegraph key, would cause the boat to respond. His control point was at the far end of the great arena.

The demonstration created a sensation and Tesla again was the popular hero. It was a front-page story in the newspapers. Everyone knew the accomplishment was a wonderful one, but few grasped the significance of the event or the importance of the fundamental discovery which it demonstrated. The basic aspects of the invention were obscured by the glamor of the demonstration.

The Spanish American War was under way. The success of the U.S. Navy in destroying the Spanish fleets was the leading topic of conversation. There was resentment over the blowing up of the U.S.S. *Maine* in Havana Harbor. Tesla's demonstration fired the imagination of everyone because of its possibilities as a weapon in naval warfare.

Waldemar Kaempffert, then a student in City College and now Science Editor of the New York *Times,* discussed its use as a weapon with Tesla.

"I see," said Kaempffert, "how you could load an even larger boat with a cargo of dynamite, cause it to ride submerged, and explode the dynamite whenever you wished by pressing the key just as easily as you can cause the light on the bow to shine, and blow up from a distance by wireless even the largest of battleships." (Edison had earlier designed an electric torpedo which received its power by a cable that remained connected with the mother ship.)

Tesla was patriotic, and was proud of his status, which he had acquired in 1889, as a citizen of the United States. He had offered his invention to the Government as a naval weapon, but at heart he was opposed to war.

"You do not see there a wireless torpedo," snapped back Tesla with fire flashing in his eyes, "you see there the first of a race of robots, mechanical men which will do the laborious work of the human race."

The "race of robots" was another of Tesla's original and important contributions to human welfare. It was one of the items of his colossal project for increasing human energy and improving the efficiency of its utilization. He visualized the application of the robot idea to warfare as well as to peaceful pursuits; and out of the broad principles enunciated, he developed an accurate picture of warfare as it is being carried on today with the use of giant machines as weapons—the robots he described.

"This evolution," he stated in an article in the *Century Magazine* of June, 1900, "will bring more and more into prominence a machine or mechanism with the fewest individuals as an element of warfare. . . . Greatest possible speed and maximum rate of energy delivery by the war apparatus will be the main object. The loss of life will become smaller. . . ."

Outlining the experiences that led him to design the robots, or automatons, as he called them, Tesla stated:

I have by every thought and act of mine, demonstrated, and do so daily, to my absolute satisfaction that I am an automaton endowed with power of movement, which merely responds to external stimuli beating upon my sense organs, and thinks and moves accordingly. . . .

With these experiences it was only natural that, long ago, I conceived the idea of constructing an automaton which would mechanically represent me, and which would respond, as I do myself, but, of course, in a much more primitive manner, to external influences. Such an automaton evidently had to have motive power, organs for locomotion, directive organs, and one or more sensitive organs so adapted as to be excited by external stimuli.

This machine would, I reasoned, perform its movements in the manner of a living being, for it would have all of the chief elements of the same. There was still the capacity for growth, propagation, and, above all, the mind which would be wanting to make the model complete. But growth was not necessary in this case since a machine could be manufactured full-grown, so to speak. As to capacity for propagation, it could likewise be left out of consideration, for in the mechanical model it merely signified a process of manufacture.

Whether the automaton be of flesh and bone, or of wood and steel, mattered little, provided it could perform all the duties required of it like an intelligent being. To do so it would have to have an element corresponding to the mind, which would effect the control of its movements and operations, and cause it to act, in any unforeseen case that might present itself, with knowledge, reason, judgement and experience. But this element I could easily embody in it by conveying to it my own intelligence, my own understanding. So this invention was evolved, and so a new art came into existence, for which the name "telautomatics" has been suggested, which means the art of controlling the movements and operations of distant automatons.

In order to give the automaton an individual identity it would be provided with a particular electrical tuning, Tesla explained, to which it alone would respond when waves of that particular frequency were sent from a control transmitting station; and other automatons would remain inactive until their frequency was transmitted. This was Tesla's fundamental radio tuning invention, the need for which other radio inventors had not yet glimpsed although Tesla had described it publicly a half-dozen years earlier.

Tesla not only used in the control of his automaton the long waves now used in broadcasting—which are very different from the short waves used by Marconi and all others; for those could be interfered with by the imposition of an intervening object—but he was explaining the use, through his system of tuning, of the spectrum of allocations for individual stations that now appears on the dials of radio receiving sets. He continued:

By the simple means described the knowledge, experience, judgement—the mind, so to speak—of the distant operator were embodied in that machine, which was thus enabled to move and perform all of its operations with reason and intelligence. It behaved just like a blindfolded person obeying directions received through the ear.

The automatons so far constructed had "borrowed minds," so to speak, as each formed merely part of the distant operator who conveyed to it his intelligent orders; but this art is only in the beginning.

I purpose to show that, however impossible it may now seem, an automaton may be contrived which will have its "own mind," and by this I mean that it will be able, independently of any operator, left entirely to itself, to perform, in response to external influences affecting its sensitive organs, a great variety of acts and operations as if it had intelligence.

It will be able to follow a course laid out or to obey orders given far in advance; it will be capable of distinguishing between what it ought and ought not to do, and of making experiences or, otherwise stated, of recording impressions which will definitely affect its subsequent actions. In fact I have already conceived such a plan.

Although I evolved this invention many years ago and explained it to my visitors very frequently in my laboratory demonstrations, it was not until much later, long after I had perfected it, that it became known, when, naturally enough, it gave rise to much discussion and to sensational reports.

But the true significance of this new art was not grasped by the majority, nor was the great force of the underlying principle recognized. As nearly as I could judge from the numerous comments which then appeared, the results I had obtained were considered as entirely impossible. Even the few who were disposed to admit the practicability of the invention saw in it merely an automobile tor-pedo, which was to be used for the purpose of blowing up battle-ships, with doubtful success. . . .

But the art I have evolved does not contemplate merely the change of direction of a moving vessel; it affords means of absolutely con-trolling in every respect, all the innumerable translatory movements, as well as the operations of all the internal organs, no matter how many, of an individualized automaton.

Tesla, in an unpublished statement, prepared fifteen years later, recorded his experience in developing automata, and his unsuccessful effort to interest the War Department, and like-wise commercial concerns, in his wirelessly controlled devices.

The idea of constructing an automaton, to bear out my theory, presented itself to me early but I did not begin active work until 1893, when I started my wireless investigations. During the succeed-ing two or three years, a number of automatic mechanisms, actuated from a distance by wireless control, were constructed by me and exhibited to visitors in my laboratory.

In 1896, however, I designed a complete machine capable of a multitude of operations, but the consummation of my labors was delayed until later in 1897. This machine was illustrated and de-scribed in my article in the Century Magazine of June 1900, and other periodicals of that time and, when first shown in the beginning of 1898, it created a sensation such as no other invention of mine has ever produced.

In November 1898, a basic patent on the novel art was granted to me, but only after the Examiner-in-Chief had come to New York and witnessed the performance, for what I claimed seemed unbelievable. I remember that when later I called on an official in Washington, with a view of offering the invention to the Government, he burst out in laughter upon my telling him what I had accomplished. No-body thought then that there was the faintest prospect of perfecting such a device.

It is unfortunate that in this patent, following the advice of my attorneys, I indicated the control as being effected through the medium of a single circuit and a well-known form of detector, for the reason that I had not yet secured protection on my methods and apparatus for individualization. As a matter of fact, my boats were controlled through the joint action of several circuits and interference of every kind was excluded. Most generally I employed receiving circuits in the form of loops, including condensers, because the discharges of my high tension transmitter ionized the air in the hall so that even a very small aerial would draw electricity from the surrounding atmosphere for hours.

Just to give an idea, I found, for instance, that a bulb 12″ in diameter, highly exhausted, and with one single terminal to which a short wire was attached, would deliver well on to one thousand successive flashes before all charge of the air in the laboratory was neutralized. The loop form of receiver was not sensitive to such a disturbance and it is curious to note that it is becoming popular at this late date. In reality it collects much less energy than the aerials or a long grounded wire, but it so happens that it does away with a number of defects inherent to the present wireless devices.

In demonstrating my invention before audiences, the visitors were requested to ask any questions, however involved, and the automaton would answer them by signs. This was considered magic at that time but was extremely simple, for it was myself who gave the replies by means of the device.

At the same period another larger telautomatic boat was constructed. It was controlled by loops having several turns placed in the hull, which was made entirely water tight and capable of submergence. The apparatus was similar to that used in the first with the exception of certain special features I introduced as, for example, incandescent lamps which afforded a visible evidence of the proper functioning of the machine and served for other purposes.

These automata, controlled within the range of vision of the operator, were, however, the first and rather crude steps in the evolution of the Art of Telautomatics as I had conceived it. The next logical improvement was its application to automatic mechanisms beyond the limits of vision and at great distances from the center of control, and I have ever since advocated their employments as instruments of warfare in preference to guns. The importance of this now seems to be recognized, if I am to judge from casual announcements through the press of achievements which are said to be extraordinary but contain no merit of novelty whatever.

In an imperfect manner it is practicable, with the existing wireless plants, to launch an aeroplane, have it follow a certain approximate course, and perform some operation at a distance of many hundreds of miles. A machine of this kind can also be mechanically controlled in several ways and I have no doubt that it may prove

*171*

of some usefulness in war. But there are, to my best knowledge, no instrumentalities in existence today with which such an object could be accomplished in a precise manner. I have devoted years of study to this matter and have evolved means, making such and greater wonders easily realizable.

As stated on a previous occasion, when I was a student at college I conceived a flying machine quite unlike the present ones. The underlying principle was sound but could not be carried into practice for want of a prime-mover of sufficiently great activity. In recent years I have successfully solved this problem and am now planning aerial machines devoid of sustaining planes, ailerons, propellers and other external attachments, which will be capable of immense speeds and are very likely to furnish powerful arguments for peace in the near future. Such a machine, sustained and propelled entirely by reaction, can be controlled either mechanically or by wireless energy. By installing proper plants it will be practicable to project a missile of this kind into the air and drop it almost on the very spot designated which may be thousands of miles away. But we are not going to stop at this.

Tesla is here describing—nearly fifty years ago—the radio-controlled rocket, which is still a confidential development of World War II, and the rocket bombs used by the Germans to attack England. The rocket-type airship is a secret which probably died with Tesla, unless it is contained in his papers sealed by the Government at the time of his death. This, however, is unlikely, as Tesla, in order to protect his secrets, did not commit his major inventions to paper, but depended on an almost infallible memory for their preservation.

"Telautomata," he concluded, "will be ultimately produced, capable of acting as if possessed of their own intelligence and their advent will create a revolution. As early as 1898 I proposed to representatives of a large manufacturing concern the construction and public exhibition of an automobile carriage which, left to itself, would perform a great variety of operations involving something akin to judgment. But my proposal was deemed chimerical at that time and nothing came from it."

Tesla, at the Madison Square Garden demonstration in 1898

which lasted for a week, presented to the world, then, two stupendous developments, either of which alone would have been too gigantic to have been satisfactorily assimilated by the public in a single presentation. Either one of the ideas dimmed the glory of the other.

This first public demonstration of wireless, the forerunner of modern radio, in the amazing stage of development to which Tesla carried it, at this early date, was too tremendous a project to be encompassed within a single dramatization. In the hands of a competent public-relations councillor, or publicity man, as he was called in those days (but the employment of one was utterly abhorrent to Tesla), this demonstration would have been limited to the wireless aspect alone, and would have included just a simple two-way sending-and-receiving set for the transmission of messages by the Morse dots and dashes. Suitably dramatized, this would have been a sufficient thrill for one show. At a subsequent show he could have brought in the tuning demonstration which would have shown the selective response of each of a series of coils, indicated by his strange-looking vacuum-tube lamps. The whole story of just the tuning of wireless circuits and stations to each other was too big for any one demonstration. An indication of its possibilities was all the public could absorb.

The robot, or automaton, idea was a new and an equally stupendous concept, the possibilities of which were not lost, however, on clever inventors; for it brought in the era of the modern labor-saving device—the mechanization of industry on a mass-production basis.

Using the Tesla principles, John Hays Hammond, Jr. developed an electric dog, on wheels, that followed him like a live pup. It was motor operated and controlled by a light beam through selenium cells placed behind lenses used for eyes. He also operated a yacht, entirely without a crew, which was sent out to sea from Boston harbor and brought back to its wharf by wireless control.

A manless airplane was developed toward the close of the

First World War. It rose from the ground, flew one hundred miles to a selected target, dropped its bombs, and returned to its home airport, all by wireless control. It was also developed so that on a signal from a distant radio station the plane would rise into the air, choose the proper direction, fly to a city hundreds of miles away and set itself down in the airport at that city. This Tesla-type robot was developed in the plant of the Sperry Gyroscope Company, where Elmer Sperry invented a host of amazing mechanical robots controlled by gyroscopes, such as the automatic pilots for airplanes and for ships.

All of the modern control devices using electronic tubes and electric eyes that make machines seem almost human and enable them to perform with superhuman activity, dependability, accuracy and low cost, are children of Tesla's robot, or automaton. The most recent development, in personalized form, was the mechanical man, a metal human monster giant, that walked, talked, smoked a cigarette, and obeyed spoken orders, in the exhibit of the Westinghouse Electric and Manufacturing Company at the New York World's Fair. Robots have been used, as well, to operate hydroelectric powerhouses and isolated substations of powerhouses.

In presenting this superabundance of scientific discovery in a single demonstration, Tesla was manifesting the superman in an additional rôle that pleased him greatly—that of the man magnificent. He would astound the world with a superlative demonstration not only of the profundity of the accomplishments of the superman, but, in addition, of the prolific nature of the mind of the man magnificent who could shower on the world a superabundance of scientific discoveries.

## *ELEVEN*

TESLA was now ready for new worlds to conquer. After presenting to the public his discoveries relating to wireless signaling or the transmission of intelligence, as he called it, Tesla
was anxious to get busy on the power phase: his projected
world-wide distribution of power by wireless methods.

Again Tesla was faced with a financial problem—or, to state
the matter simply, he was broke. The $40,000 which was paid
for the stock of the Nikola Tesla Company by Adams had been
spent. The company had no cash on hand; but it held patents
worth many millions if they had been handled in a practical
way. A gift of $10,000 from John Hays Hammond, the famous
mining engineer, had financed the work leading up to the Madison Square Garden wireless and robot demonstration.

Tesla had built ever larger and more powerful oscillators in
his Houston Street laboratory. When he constructed one that
produced 4,000,000 volts he reached beyond the limits in which
high voltages could be handled within a city building. The
sparks jumped to the walls, floors and ceilings. He needed a
larger open space. He wanted to build vastly larger coils. He
dreamed of a tremendous structure he would like to build somewhere in the open country spaces. He felt certain his wireless
patents would prove tremendously valuable in a short time, and
he would then have all the money he needed to build his laboratory. But he had already progressed to the point at which further
advancement demanded the use of such a building—and he was
broke. A loan of $10,000 offered by his friend Crawford, of the
dry goods firm of Simpson and Crawford, took care of immediate
needs.

Leonard E. Curtis, of the Colorado Springs Electric Company,
a great admirer of Tesla, when he heard of Tesla's plan to con-

duct experiments on a gigantic scale, invited him to locate his laboratory at Colorado Springs, where he would provide him with the necessary land and all the electric power he needed for his work.

Col. John Jacob Astor, owner of the Waldorf-Astoria, held his famous dining-room guest in the highest esteem as a personal friend, and kept in close touch with the progress of his investigations. When he heard that his researches were being halted through lack of funds, he made available to Tesla the $30,000 he needed in order to take advantage of Curtis' offer and build a temporary plant at Colorado Springs. Tesla arrived in Colorado in May, 1899, bringing with him some of his laboratory workers, and accompanied by an engineering associate, Fritz Lowenstein.

While Tesla was making experiments on natural lightning and other subjects in his mountain laboratory, the construction work on his high-power transmitting apparatus was being rushed. He gave his personal supervision to even the finest details of every piece of apparatus. He was working in a virgin field. None had gone before him to pave the way or gain experience that would be helpful to him in designing his experiments or his machines. He was entirely on his own, working without human guidance of any kind, exploring a field of knowledge far beyond that which anyone else had reached. He had previously astounded the world in developing a system of power transmission in which pressures of tens of thousands of volts were used; now he was working with millions of volts, and no one knew what would happen when such tremendous potentials were produced. He believed, however, that he would make his own magnificent polyphase system obsolete by creating a better one.

In about three months after his arrival at Colorado Springs the building with its fantastic shapes, towers and masts was completed, and the giant oscillator with which the principal experiment was to be made was ready for operation.

The wild, rugged, mountainous terrain of Colorado, in which

Tesla set up his laboratory, is a natural generator of tremendous electrical activity, producing lightning discharges of a magnitude and intensity probably not equaled anywhere else on earth. Overwhelming bolts from both earth and sky flashed with frightening frequency during the almost daily lightning storms. Tesla made a very detailed study of natural lightning while his apparatus, which would imitate it, was being constructed. He learned a great deal about the characteristics of the various kinds of discharges.

The gods of the natural lightning may have become a bit jealous of this individual who was undertaking to steal their thunder, as Prometheus had stolen fire, and sought to punish him by wrecking his fantastic looking structure. It was badly damaged, and narrowly escaped destruction, by a bolt of lightning, not one that made a direct hit but one that struck ten miles away.

The blast hit the laboratory at the exact time, to the split second, that Tesla predicted it would. It was caused by a tidal wave of air coming from a particular type of lightning discharge. Tesla tells the story in an unpublished report. He stated:

I have had many opportunities for checking this value by observation of explosions and lightning discharges. An ideal case of this kind presented itself at Colorado Springs in July 1899 while I was carring on tests with my broadcasting power station which was the only wireless plant in existence at that time.

A heavy cloud had gathered over Pikes Peak range and suddenly lightning struck at a point just ten miles away. I timed the flash instantly and upon making a quick computation told my assistants that the tidal wave would arrive in 48.5 seconds. Exactly with the lapse of this time interval a terrific blow struck the building which might have been thrown off the foundation had it not been strongly braced. All the windows on one side and a door were demolished and much damage done in the interior.

Taking into account the energy of the electric discharge and its duration, as well as that of an explosion, I estimated that the concussion was about equivalent to that which might have been produced at that distance by the ignition of twelve tons of dynamite.

The experimental station which Tesla erected was an almost square barnlike structure nearly one hundred feet on each side. The sides were twenty-five feet high, and from them the roof sloped upward toward the center. From the middle of the roof rose a skeleton pyramidal tower made of wood. The top of this tower was nearly eighty feet above the ground. Extensions of the slanting roof beams extended outward to the ground to serve as flying buttresses to reinforce the tower. Through the center of the tower extended a mast nearly two hundred feet high, at the top of which was mounted a copper ball about three feet in diameter. The mast carried a heavy wire connecting the ball with the apparatus in the laboratory. The mast was arranged in sections so that it could be disjointed and lowered.

There were many pieces of apparatus in the building, and many forms and sizes of his Tesla coils, or high-frequency current transformers. The principal device was his "magnifying transmitter." This was merely a very large Tesla coil. A circular fence-like wall seventy-five feet in diameter was built in the large central room of the structure, and on this were wound the turns of the giant primary coil of the magnifying transmitter. The secondary was a coil about ten feet in diameter, of about seventy-five turns of wire wound on a cylindrical skeletonized framework of wood. It had a vertical length of about ten feet and was mounted in the center of the room several feet above the floor. In the center of this coil was the bottom part of the mast. The roof above this portion of the room could be slid outward in two sections, so that no material came within a long distance of the mast and its wire conductor within the lower third of the distance above the ground.

One of the first problems Tesla sought to solve when he began his researches in the mountains of Colorado was whether the earth was an electrically charged body. Nature is usually very generous in her response when scientists ask her, in their experiments, questions of first magnitude. Tesla not only received a very satisfactory answer to his question but in addition a revelation of tremendous importance, an unveiling of a secret of Nature's operations which places in the hands of man a means of manipulating electrical forces on a terrestrial scale.

It was desirable for Telsa to learn whether the earth was electrically charged for the same reason that a violinist would want to know whether the strings of his instrument lay loose and inert across the bridge or whether they were tense and taut so that they would produce a musical note if plucked, or a football player would want to know if the pigskin were inflated or limp.

If the earth were uncharged, it would act as a vast sink into which electricity would have to be flowed in tremendous amounts to bring it to the state in which it could be made to vibrate electrically. An uncharged earth would somewhat complicate Tesla's plans. He quickly discovered that the earth is charged to an extremely high potential and is provided with some kind of a mechanism for maintaining its voltage. It was while determining this fact that he made his second big discovery.

Tesla made the first announcement of his discovery shortly after his return to New York in an amazing article in the *Century* of June, 1900, but the story is best told by Tesla in an article in the *Electrical World and Engineer*, May 5, 1904:

In the middle of June, while preparations for other work were going on, I arranged one of my receiving transformers with the view of determining in a novel manner, experimentally, the electrical potential of the globe and studying its periodic and casual fluctuations. This formed part of a plan carefully mapped out in advance.

A highly sensitive, self restorative device, controlling a recording instrument, was included in the secondary circuit, while the primary was connected to the ground and the secondary to an elevated terminal of adjustable capacity. The variations of electrical potential gave rise to electrical surgings in the primary; these generated secondary currents, which in turn affected the sensitive device and recorder in proportion to their intensity.

The earth was found to be, literally, alive with electrical vibrations, and soon I was deeply absorbed in this interesting investigation. No better opportunity for such observations as I intended to make could be found anywhere.

Colorado is a country famous for the natural displays of electric force. In that dry and rarefied atmosphere the sun's rays beat on objects with fierce intensity. I raised steam to a dangerous pressure, in barrels filled with concentrated salt solution and the tinfoil coating of some of my elevated terminals shrivelled in the fiery blaze. An experimental high tension transformer, carelessly exposed to the rays of the setting sun, had most of its insulating compound melted and was rendered useless.

Aided by the dryness and rarefaction of the air, the water evaporates as in a boiler, and static electricity is generated in abundance. Lightning discharges are, accordingly, very frequent and sometimes of inconceivable violence. On one occasion approximately 12,000 discharges occurred within two hours, and all in the radius of certainly less than 50 kilometers [about 30 miles] from the laboratory. Many of them resembled gigantic trees of fire with the trunks up or down. I never saw fireballs, but as a compensation for my disappointment I succeeded later in determining the mode of their formation and producing them artificially.

In the latter part of the same month I noticed several times that my instruments were affected stronger by discharges taking place at great distances than by those near by. This puzzled me very much. What was the cause? A number of observations proved that it could not be due to differences in the intensity of individual discharges and I readily ascertained that the phenomenon was not the result of a varying relation between the periods of my receiving circuits and those of the terrestrial disturbances.

One night as I was walking home with an assistant, meditating over these experiences, I was suddenly staggered by a thought. Years ago when I wrote a chapter of my lecture before the Franklin Institute and the National Electric Light Association, it had presented itself to me, but I dismissed it as absurd and impossible. I banished it again. Nevertheless, my instinct was aroused and somehow I felt that I was nearing a great revelation.

It was on the 3rd of July [1899]—the date I shall never forget—when I obtained the first decisive experimental evidence of a truth of overwhelming importance for the advancement of humanity.

A dense mass of strongly charged clouds gathered in the west and toward evening a violent storm broke loose which, after spending much of its fury in the mountains, was driven away with great velocity over the plains. Heavy and long persistent arcs formed almost in regular time intervals. My observations were now greatly facilitated and rendered more accurate by the experiences already gained. I was able to handle my instruments quickly and I was prepared. The recording apparatus being properly adjusted, its indications became fainter and fainter with increasing distance of the storm, until they ceased altogether.

I was watching in eager expectation. Surely enough, in a little while the indications again began, grew stronger and stronger, and, after passing through a maximum, gradually decreased and ceased once more. Many times, in regularly recurring intervals, the same actions were repeated until the storm which, as evident from simple computations, was moving with nearly constant speed, had retreated to a distance of about 300 kilometers [about 180 miles]. Nor did these strange actions stop then, but continued to manifest themselves with undiminished force.

Subsequently similar observations were also made by my assistant, Mr. Fritz Lowenstein, and shortly afterward several admirable opportunities presented themselves which brought out, still more forcibly and unmistakably, the true nature of the wonderful phenomenon. No doubt whatever remained: I was observing stationary waves.

As the source of the disturbances moved away the receiving circuit came successively upon their nodes and loops. Impossible as it seemed, this planet, despite its vast extent, behaved like a conductor of limited dimensions. The tremendous significance of this fact in the transmission of energy by my system had already become quite clear to me.

Not only was it practicable to send telegraphic messages to any distance without wires, as I recognized long ago, but also to impress upon the entire globe the faint modulations of the human voice, far more still, to transmit power, in unlimited amounts to any terrestrial distance and almost without loss.

To get a more familiar picture of the problem that Tesla tackled in seeking to determine if the earth were charged and if it could be set into electrical vibration, one can visualize the difference between a bath tub that is empty and one that contains water. The uncharged earth would be like an empty tub; the charged earth like one containing water. It is easy to produce waves in the tub containing the water. By placing one's hand in the water and moving it back and forth, lengthwise, a short distance at the right rhythm, the water is soon rushing back and forth in a wave whose amplitude grows at a tremendously rapid rate until, if the hand motion is continued, the water may splash as high as the ceiling.

The earth can be visualized as an extremely large container holding a fluid; and in the center is a small plunger arrangement which can be moved up and down a short distance in the proper rhythm. The waves travel to the edge of the container and are reflected back to the center, from which they again go outward re-enforced by the movement of the plunger.

The reaction between the outgoing and incoming waves, both in resonance with the medium in which they are traveling, causes stationary waves to be produced on the water, the surface having the appearance of a single series of waves frozen in a fixed position.

In Tesla's experiments the lightning discharges that played the part of the plunger causing the waves were moving rapidly to the eastward, and they carried the whole series of fixed, or stationary, waves with them. The measuring device remained fixed so the wave series, with its loops and nodes, moved past it, causing the measured potentials to rise and fall.

The experiment not only demonstrated that the earth was filled with electricity, but that this electricity could be disturbed so that rhythmic vibrations could be struck, resonance could be produced, causing effects of tremendous magnitude. Soldiers marching in unison across a bridge and wrecking it by the resulting vibration would again be a case in point.

Tesla produced the spectacular effects of extremely high potentials and high frequency by producing electrical resonance in his circuits—by tuning the electricity—and now he had discovered that he would be able to produce, easily, the same effect in the earth as if it were a single condenser and coil combined, a pure electrical resonating unit, by charging and discharging it rhythmically with his high-frequency, high-potential oscillations.

In this magnificent experiment Tesla the superman was at his best; the boldness of his undertaking fired the imagination, and the success he achieved should have earned for him undying fame.

EVENTUALLY, the giant coils with their banks of condensers and other apparatus in the Colorado laboratory were ready for use in full-scale experiments. Every piece of equipment was thoroughly inspected and tested by Tesla—and the moment had finally arrived for the critical test of the highest voltage experiment that had ever been made. He expected to top his own earlier records one hundred times over, and to produce tens of thousands of times higher voltages than ever had been produced in the high-voltage transmission lines at Niagara Falls.

There was not the faintest shadow of doubt in Tesla's mind as to whether his giant oscillator would work. He knew it would work, but he also knew that he was going to produce millions of volts and tremendously heavy currents; and neither he nor anyone else knew how these terrific explosions of electrical energy would act. He knew that he had planned the experiment so that the first bolts of man-made lightning ever created would shoot from the top of the 200-foot-tall mast.

Tesla asked Kolman Czito, who had worked with him for many years in his laboratories in New York, to preside at the switchboard through which current was brought into the laboratory from the powerhouse of the Colorado Springs Electric Company by an overhead transmission line two miles long.

"When I give you the word," said Tesla to Czito, "you close the switch for one second—not longer."

The inventor took a position near the door of the laboratory from which he could view the giant coil in the center of the great barnlike room—but not too close to it, for a stray bolt of his own lightning might inflict a painful burn. From the point where he stood he could look upward toward the open roof and see the three-foot copper ball on top of the slender 200-foot mast that had its base in the center of the cagelike secondary coil. A quick visual survey of the situation, and Tesla gave the signal— "Now."

Czito jammed home the switch and as quickly pulled it out. In that brief interval the secondary coil was crowned with a mass of hairlike electrical fire, there was a crackling sound in various parts of the room and a sharp snap far overhead.

"Fine," said Tesla, "the experiment is working beautifully. We will try it again in exactly the same way. Now!"

Again Czito jammed home the switch for a second and opened it. Again the plumes of electrical fire came from the coil, minor sparks crackled in all parts of the laboratory and the very sharp snap came through the open roof from far overhead.

"This time," said Tesla, "I am going to watch the top of the mast from the outside. When I give you the signal I want you to close the switch and leave it closed until I give you the signal to open it." So saying, he started for the near-by open door.

Reaching a point outside from which he could see the copper ball on top of the needlelike mast, Tesla called through the door, "Czito, close the switch— Now!"

Czito again jammed the switch closed and jumped back—but held his arm extended to yank open the blades quickly should he receive an emergency signal. Nothing much had happened on the quick-contact closing, but now the apparatus would be given an opportunity to build up its full strength and no one knew what to expect. He knew that the apparatus would draw a very heavy current through a primary coil that looked like a "short circuit," and he knew that short circuits could be very destructive if the current was allowed to continue to flow. The switchboard could become a scene of interesting activity if any-thing let go. Czito expected the quick flash and explosive blast of a short circuit a second or two after the switch was closed. Several seconds passed with no short circuit.

As soon as the switch was closed there came again the same

crackling sound, the same snap high in the air that he had heard before. Now it was followed by a tremendous upsurge of sound. The crackling from the coil swelled into a crescendo of vicious snaps. From above the roof the original staccato snap was followed by a sharper one—and by another that was like the report of a rifle. The next was still louder. They came closer together like the rattle of a machine gun. The bang high in the air became tremendously louder; it was now the roar of a cannon, with the discharges rapidly following each other as if a gigantic artillery battle was taking place over the building. The sound was terrifying and the thunder shook the building in most threatening fashion.

There was a strange ghostly blue light in the great barnlike structure. The coils were flaming with masses of fiery hair. Everything in the building was spouting needles of flame, and the place filled with the sulphurous odor of ozone, fumes of the sparks, which was all that was needed to complete the conviction that hell was breaking loose and belching into the building.

As Czito stood near the switch he could feel and see the sparks jump from his fingers, each pricking like a needle stuck into his flesh. He wondered if he would be able to reach for the switch and turn off the power that was creating this electrical pandemonium—would the sparks become longer and more powerful if he approached the switch? Must this head-splitting racket go on forever? It's getting worse, that tremendous ear-wrecking bang, bang, bang overhead. Why doesn't Tesla stop it before it shakes down the building? Should he open the switch of his own accord? Maybe Tesla has been hit, perhaps killed, and can't give orders to open the switch!

It seemed to Czito that the demonstration had been going on for an hour but as a matter of fact it had lasted thus far for only a minute; nevertheless, a tremendous amount of activity had been crowded into that short space of time.

Outside stood Tesla, properly attired in cutaway coat and black derby hat for the auspicious occasion, his slender six-foot-two figure bearing signs of close relationship to the mastlike rod sticking out of his bizarre barnlike structure. His height was increased by a one-inch-thick layer of rubber on the soles and heels of his shoes, used as electrical insulation.

As he gave the switch-closing "Now" signal to Czito, he turned his eyes heavenward to the ball on top of the mast. He had hardly spoken when he saw a short hairlike spark dart from the ball. It was only about ten feet long, and thin. Before he had time to be pleased, there was a second and a third and a fourth spark, each longer, brighter and bluer than its predecessor.

"Ah!" ejaculated Tesla, forgetting to close his mouth that was widely opened for a shout. He clenched his hands for joy and raised them skyward toward the top of the mast.

More sparks! Longer and longer! Ten, twenty, thirty, forty, fifty, sixty, seventy, eighty feet. Brighter and bluer! Not thread-like sparks now but fingers of fire. Wriggling rods of flame that lashed viciously into the heavens. The sparks were now as thick as his arm as they left the ball.

Tesla's eyes almost popped out of his head as he saw full-fledged bolts of lightning darting into the air, accompanied by a barrage of tremendous crashes of thunder. Those lightning bolts were now half again the length of the building, more than 135 feet long, and the thunder was being heard in Cripple Creek fifteen miles away.

Suddenly—silence!

Tesla rushed into the building.

"Czito! Czito! Czito! Why did you do that? I did not tell you to open the switch. Close it again quickly!"

Czito pointed at the switch. It was still closed. He then pointed at the voltmeter and ammeter on the switchboard. The needles of both of them registered zero.

Tesla sized up the situation instantly. The incoming wires carrying power to the laboratory were "dead."

"Czito," he snapped, "call up the powerhouse quickly. They must not do that. They have cut off my power."

The telephone call was put through to the powerhouse. Tesla grabbed the phone and shouted into it:

"This is Nikola Tesla. You have cut off my power! You must give me back power immediately! You must not cut off my power."

"Cut off your power, nothing," came the gruff reply from the other end of the line. "You've thrown a short circuit on our line with your blankety-blank-blank experiments and wrecked our station. You've knocked our generator off the line and she's now on fire. You won't get any more power!"

Tesla had built his apparatus substantially, so that it would be able to carry the tremendously heavy currents he expected to draw off the line. While his own equipment was able to stand what amounted to a heavy short circuit, he had overloaded the generator at the Colorado Springs Electric Company powerhouse, which tried manfully to carry the added burden—but the heavy surge of current was too much for the dynamo that was not designed to stand such heavy overloads. Its wires became hotter and hotter, and finally the insulation took fire and the copper wire in the armature coils melted like wax, opening its circuits so that it ceased to generate electricity.

The powerhouse had a second, standby, generator which was started up in a short time. Tesla was insistent that he be supplied with current from this machine as soon as it was running, but his demand was refused. In the future, he was told, he would be supplied with current from a dynamo operated independently from the one supplying the company's regular customers. The independent dynamo, he was told, would be the one that was already burned out—and he would get no service until it was repaired. Tesla offered to pay the cost of an extra-special rush job on the repairs if he were permitted to handle the work. Alternating-current dynamos were no mystery to him. Taking his workers from the laboratory to the powerhouse he soon had the repair job under way, and in less than a week the dynamo was again operating.

A LIGHTNING stroke produces its spectacular pyrotechnics and earth-vibrating effects with less than a nickel's worth of electricity—at a five-cent-a-kilowatt hour rate, which is somewhat less than the average household rate for current. It consists of tremendously heavy currents, many thousands of amperes at millions of volts, but it lasts only a few millionths of a second. If supplied with this "nickel's worth" of current continuously, the lightning flash would last indefinitely.

Tesla, in his Colorado Springs laboratory, was pumping a steady flow of current worth, at the above rate, about $15.00 an hour into the earth. In an hour he charged the earth with several hundred times as much electrical energy as is contained in a single lightning stroke. Owing to resonance phenomena, he could build up electrical effects in the earth greatly exceeding those of lightning since it was only necessary, once resonance was established, to supply energy equal to frictional losses, in order to maintain this condition.

In describing his work with the giant oscillator, Tesla, using conservative estimates of his results, stated in his article in the *Century Magazine* of June, 1900:

> However extraordinary the results shown may appear, they are but trifling compared with those attainable by apparatus designed on these same principles. I have produced electrical discharges the actual path of which, from end to end, was probably more than 100 feet long; but it would not be difficult to reach lengths 100 times as great.
>
> I have produced electrical movements occurring at the rate of approximately 100,000 horsepower, but rates of one, five or ten million horsepower are easily practicable. In these experiments effects were developed incomparably greater than ever produced by any human agencies, and yet these results are but an embryo of what is to be.

The method used by Tesla to set the earth in electrical oscillation is the electrical counterpart of the mechanical device previously described, the plunger bobbing up and down at the right rhythm that created the stationary waves in the water.

Tesla used a stream of electrons which were pumped into and

drawn out of the earth at a rapid rhythmic rate. At the time the experiments were made, the electron still was not known to be the fundamental atom of electricity, so the operation was spoken of simply as the flow of electricity.

The pumping operation was carried on at a rate of 150,000 oscillations per second. These would produce electrical pulsations with a wavelength of 2,000 meters (about 6,600 feet).

When the moving waves expanded outward from Colorado Springs, they traveled in all directions in ever increasing circles until they passed over the bulge of the earth, and then in ever smaller circles and with increasing intensity converged on the diametrically opposite point of the earth, a trifle to the west of the two French Islands, Amsterdam and St. Paul, in the area between the Indian and Antarctic Oceans midway between the southern tip of Africa and the southwest corner of Australia. Here a tremendous electrical south pole was built up, marked by a wave of great amplitude that rose and fell in unison with Tesla's apparatus at its north pole in Colorado Springs. As this wave fell, it sent back an electrical echo which produced the same effect at Colorado Springs. Just as it arrived back at Colorado Springs, the oscillator was working to build up a wave that would re-enforce it and send it back more powerfully than before to the antipode to repeat the performance.

If there were no losses in this operation—if the earth were a perfect electrical conductor, and there were no other sources of resistance—this resonance phenomenon would build up to a destructive action of gigantic proportions, even with the charging source of only about 300 horsepower which Tesla used. Voltages of gigantic magnitudes would be built up. Charged particles of matter would be hurled outward from the earth with vast energies, and eventually even the solid matter of the earth would be affected and the whole planet disintegrated. Pure resonance, however, is not attainable. Tesla frequently stressed the fortunate nature of this fact; for otherwise disastrous results could be produced by small amounts of energy. The electrical resistance of the globe would prevent the attainment of pure resonance; but practical resonance can be attained with safety by supplying continuously the amount of energy lost in resistance—and this supplies perfect control of the situation.

With the earth set in electrical oscillation, a source of energy is provided at all spots on the earth. This could be drawn off and made available for use by a suitable simple apparatus which would contain the same elements as the tuning unit in a radio set, but larger (a coil and a condenser), a ground connection and metal rod as high as a cottage. Such a combination would absorb, at any point on the earth's surface, energy from the waves rushing back and forth between the electrical north and south poles created by the Tesla oscillators. No other equipment would be needed to supply light to the home, provided with Tesla's simple vacuum-tube lamps, or to produce heating effects.*

The apparatus that Tesla used to charge the earth is very simple in principle. In its elementary form it consists of a circuit containing a large coil and condenser of the correct electrical dimensions to give it the desired frequency of oscillation, a source of electric current for energizing the circuit, and a step-up transformer, also tuned, for increasing the voltage.

The current of a few hundred volts obtained from the powerhouse was stepped up by an ordinary iron box transformer to more than 30,000 volts and at this potential was fed into a condenser which, when filled, discharged into the coil connected across its terminals. The rate of the back-and-forth surge of current from condenser into coil and coil back to condenser, in endless repetition, is determined by the capacity of the condenser for holding current and the length, or inductance, of the coil through which the discharge must travel. An arc between the joint

* For the operation of ordinary-type motors, a frequency changer would be needed. Tesla, indeed, developed ironless motors that would operate on high-frequency currents, but they could not compete in efficiency with motors operated on low-frequency currents. Frequency transformation, however, is now a very practical operation.

terminals of condenser and coil completed the free oscillating path of the high-frequency current.

In an oscillating circuit the current is at zero value at the start of each cycle, rises to a high value and drops to zero again at the end of each half cycle. The voltage does the same. Both build up to high values at the midpoint of each half cycle.

The coil through which the current flows is surrounded by a magnetic field produced by the current. With heavy current flows, these fields can become very extensive and of high intensity, particularly at the midpoint in each half cycle.

The primary coil, or energizing circuit of Tesla's oscillator, consisted of a number of turns of heavy wire mounted on a circular fence eighty feet in diameter in the great hall of his laboratory. In the space within this fenced enclosure the magnetic field built up to a crescendo of intensity with each half cycle of the current in the primary coil. As the magnetic circles of force moved to the center of the enclosure, they became more concentrated and built up a high density of energy in space in this region.

Centered in this area was another coil perfectly tuned to vibrate electrically in resonance with the crescendo of energy in which it was immersed 300,000 times per second. This coil—about ten feet in diameter, consisting of nearly one hundred turns on a cagelike frame about ten feet high—in responding resonantly, built up potentials with maximum values of more than 100,000,000 volts. No scientist has ever succeeded in building up currents with even one tenth of this potential since that time.

When the first surge of magnetic energy crashed into this coil, it caused a downward avalanche of electrons from the coil into the earth, thereby inflating the earth electrically and raising its potential. The next surge of magnetic energy was of the opposite polarity and caused a tidal wave of electrons from the earth to rush through the coil and upward to the terminal of the coil, which was the metal ball mounted on the mast 200 feet high.

The downward flood of electrons was spread over the wide area of the earth but the return upward flood was concentrated on a small metal ball on top of the mast, upon which tremendously high potentials developed. The electrons on the ball were under explosive electrical pressure and were forced to escape. They made a spearpoint of attack on the surrounding air, broke a small opening, and through this rushed uncounted billions of billions of electrons, their mad stampede rendering their path through the air incandescent for a distance of scores of feet—in other words, producing a flash of lightning.

Having thus succeeded in making the earth oscillate as if it were a piece of laboratory apparatus, Tesla would now proceed to test the practical applications of his unique method of world-wide power transmission.*

THE full story of Tesla's accomplishments at Colorado Springs has never been told and never will be told. He carried the records, engraved on his almost infallible memory, with him when he died. Fritz Lowenstein, a competent electrical engineer, interested in high-frequency currents, was his assistant at Colorado Springs. Tesla, however, took neither Lowenstein nor anyone else into his confidence.

It was not necessary for Tesla to write the detailed records of experiments which scientists and engineers make, as routine, of their laboratory tests. He possessed a most remarkable memory,

---

* In describing the mode of transmission of his oscillating currents through the earth, Tesla claimed the path of the discharge was from his station directly through the center of the earth and in a straight line to the antipode, the return being by the same route, and that the current on this straight-line path traveled at its normal velocity—the speed of light. This flow, he declared, produced an accompanying surface flow of current, which was in step at the starting point and when they rejoined at the antipode; and this necessitated higher velocities in flowing over the surface of the earth. The surface velocities would be infinite at each of the antipodes, and would decrease rapidly until at the equatorial region of this axis it would travel at the normal velocity of the currents.

This theory is probably incorrect, the flow probably being confined to the surface, and taking place at the normal velocity of the electric current; but this, however, had no effect on the practical operation of Tesla's oscillators in producing their earth-charging effect.

supplemented by his strange power of visualizing again, in their full aspects of reality, any past events. He needed no reference books, for he could quickly derive any desired formula from basic concepts; and he even carried a table of logarithms in his head. For these reasons there is a great lack of written records on his experiments, and what is recorded is mostly of a minor nature.

Fundamental facts of great importance that he intended to develop later in a practical manner were stored in the archives of his mind to await the time when he would be able to present a practical working model of the inventions based on his discoveries. He had no fear that he would be anticipated by others because he was so far in advance of his contemporaries that he could safely bide his time for developing his ideas.

It was Tesla's intention to make the development of his discoveries a one-man job. He was completely confident, at this time, of his ability to live a century and a quarter, and to be actively engaged in creative experimental work up to at least his one-hundredth birthday, at which time he would give serious thought to the task of writing his biography and a complete record of his experimental work. Up to almost his eightieth year he adhered to this plan without doubt as to its ultimate consummation.

As a result of this most unfortunate design, technical details are lacking concerning the principal discoveries made at Colorado Springs. By piecing together the fragmentary material published in a number of publications, however, it appears evident that Tesla, in addition to experiments with his gigantic current movements, as a means of establishing world-wide broadcasts and making a number of detectors for such use, tested his power-transmission system at a distance of twenty-six miles from his laboratory and was able to light two hundred incandescent lamps, of the Edison type, with electrical energy extracted from the earth while his oscillator was operating. These lamps consumed about fifty watts each; and if two hundred were used in the test bank, the energy consumed would be 10,000 watts, or approximately thirteen horsepower.

Transmission of thirteen horsepower wirelessly through the earth for a distance of twenty-six miles can be accepted as a very adequate demonstration of the practicability of Tesla's plan. He claimed an efficiency of higher than 95 per cent for this method of energy transmission; so he could, undoubtedly, with a 300 horsepower oscillator, operate more than a dozen such test demonstrations simultaneously anywhere on the globe. With respect to the latter point he stated, "In this new system it matters little—in fact, almost nothing—whether the transmission is effected at a distance of a few miles, or of a few thousand miles."

"While I have not as yet," he stated in the *Century* article of June, 1900, "actually effected a transmission of a considerable amount of energy, such as would be of industrial importance, to a great distance by this new method, I have operated several model plants under exactly the same conditions which will exist in a large plant of this kind, and the practicability of the system is thoroughly demonstrated."

Tesla was insistent, in his latter decades, on the existence, the actuality, the importance and availability of many undisclosed discoveries which he made at Colorado Springs. The author urged upon Tesla two or three times the desirability of making a disclosure, against the ever present danger of an accident that might cause them to be lost to the world; and when the inventor was unimpressed by this possibility, he was asked to permit the author to do something that would bring about their practical development. Tesla was courteously appreciative of the interest manifested, but he was very emphatic in his insistence that he would handle his own affairs as he saw fit, and that he expected shortly to have adequate funds to develop his inventions.

TESLA returned to New York, in the fall of 1899, broke once more, but with the knowledge that his efforts had greatly enriched humanity with important scientific discoveries. Yet even more important was the new attitude his work had made possible: man had achieved a method through which he could control his gigantic planet, could look upon this heavenly body from the godlike vantage point in which he could view it as a piece of laboratory apparatus to be manipulated as he willed.

The pictures which Tesla brought back to New York showing the gigantic electrical discharges from his oscillator, and the stories he related of his experiences, created a tremendous impression in his circle of friends. It was then that Robert Underwood Johnson, one of the editors of the *Century Magazine*, at whose home in Madison Avenue, in the exclusive Murray Hill section, Tesla was a frequent and informal visitor, requested the inventor to write an article telling of his accomplishments.

When the article was written, Johnson returned it, telling Tesla he had served a mess of cold philosophical stones instead of a dish of hot throbbing facts. The inventor had made but scant reference to his recent astounding accomplishments, but developed instead a philosophical system in which the progress of humanity was viewed as purely a mechanical process, activated by the sources of energy available. Three times the article went back to Tesla and was as many times rewritten, despite the high literary quality of the work on each occasion.

The article, which carried the title "The Problem of Increasing Human Energy," created a sensation. Among those whose interest it aroused was J. Pierpont Morgan—a most fortunate circumstance for Tesla. The great financier had a soft spot for geniuses, and Tesla was a perfect example of the species.

Morgan the financier was famous, but Morgan the philanthropist, a greater personality, was to the general public non-existent, so carefully guarded against publicity were his benefactions. In this he was not always completely successful for there are, of necessity, two parties to a beneficence, the giver and the receiver; and the pride and gratitude of the latter can develop into a weak spot in the shell of secrecy.

Tesla was invited to Morgan's home and quickly became a

favorite with the family. His record of accomplishment which promised still greater achievements in the future, his pleasant personality, his high moral standards of conduct, his celibate manner of life and his manner of subordinating himself to his work, his boylike enthusiasm, were factors that caused him to be admired not only by Morgan but by all others who knew him well.

Morgan made inquiries of Tesla concerning his financial structure. There were, in those days, a limited number of strong financial groups who were playing a terrestrial game of chess with the world's economic resources; the discoveries of a genius like Tesla might well have a profound effect on the destinies of one or more of these groups, and it would be well for an operator in this field to know more of the inventor's commitments. Undoubtedly, it was a source of surprise and satisfaction to Morgan when he learned that Tesla was a lone operator and now entirely without funds needed to carry on his researches.

Morgan knew well the inestimable value of Tesla's polyphase alternating-current system. The Niagara development was a Morgan enterprise, and gigantic plans were being builded on its already proven success. The man who laid the scientific and engineering foundation for this new and profitable industrial electrical era was broke and engaged in developing a new source of power distribution. He had supplanted Edison's half-mile power pygmy with a giant having a thousand-mile range, and now he was working on a system which experiments had shown could distribute power wirelessly to the ends of the earth with but a small fraction of the losses of the Edison system in distributing power by wire for half a mile, and could even send current around the earth cheaper than his own alternating-current system could distribute it at a distance of one hundred miles. The economic implications of this development staggered the imagination. What effect would it have on the chess game being played by the world's financial groups?

Would the new wireless-distribution-of-power system fit into the existing economic and financial structure? Could it be usefully applied without derangements of greater magnitude than the benefits it would produce? If it were adopted for development, who would be best suited to control it? Could it be controlled in a practical way when any spot on earth would be an outlet for an unlimited reservoir of power for anyone who cared to tap it with a simple device? How could compensation be collected for the service rendered?

These were some of the most obvious aspects of Tesla's world power system that would instantly present themselves to the practical mind of Morgan. In addition, Tesla was proposing a world-wide broadcasting system for distributing news, entertainment, knowledge, and a host of other interesting items. Morgan could well understand the practical aspects of wireless communication in which a charge could be made for transmitting messages from point to point, which was a part of the Tesla system—but, to the inventor's way of thinking, only a minor part compared to the more important broadcasting and power-distribution systems.

A Morgan would understand that ingenious minds could work out some method for placing such world-wide services on a practical profit-paying basis; but this whole new Tesla development had a fantastical aspect that was upsetting to so-called "practical" minds not accustomed to thinking first-magnitude thoughts. The new system might prove more important than the polyphase system which went as a record-breaking bargain to Westinghouse for $1,000,000. Westinghouse was then the most powerful competitor of the Edison system which Morgan had backed, and particularly of the General Electric Company whose financing Morgan had arranged. Although Westinghouse secured a monopoly, means were found for causing him to share it, by a license agreement, with the General Electric Company, so the Morgan company had equal opportunities to exploit the rich market.

History might now be repeating itself with the same inventor, who now had a hypersuperpower system to supplant his own superpower system. In this case Morgan could place himself in a position to seize the monopoly of world power.

The group holding a monopoly control over such a system could develop it, or not develop it, as it saw fit; it could be developed to produce a profit by supplanting or supplementing the satisfactory wire distribution system, or it could be put on the shelf to prevent it from interfering with the existing system. A monopoly of it could prevent any other group from securing it and using it as a club to force concessions from those controlling existing enterprises. Ownership of the Tesla world-power and world-broadcasting patents might well prove an extremely profit-able investment even if a very high price were paid for them.

But there was also a more subtle viewpoint. Without a strong backing by a powerful source of capital, a world-wide system such as Tesla proposed could never be brought into operating exist-ence. If a powerful group had an opportunity to get in on the ground floor and secure monopoly control and failed to do so, and let it become apparent that this was done intentionally, the effect of such a decision could easily result in scaring off any other groups and effectually preventing anyone from ever backing the system.

Morgan, however, in his contacts with Tesla, brought no commercial or practical aspects into the situation. His interest was entirely that of a patron seeking to aid a genius to express his creative talents. He made gifts to Tesla to which there were no strings attached. The inventor could use the money as he saw fit. No definite information is available as to the amount of those contributions, but an authoritative source, close to Tesla, fixes the amount which he received within a very short period at $150,000. Later contributions, spread over a long period of years, are believed to have brought the total to double this amount.

Tesla made no secret of Morgan's support. He stated, in the

article in the *Electrical World and Engineer,* published March 5, 1904, describing his wireless power work up to that time:

"For a large part of the work which I have done so far I am indebted to the noble generosity of Mr. J. Pierpont Morgan, which was all the more welcome and stimulating, as it was extended at a time when those, who have since promised most, were the greatest of doubters."

When Morgan made his first contribution, the rumor got into circulation that he was financially interested in the enterprise upon which Tesla now embarked. The resulting situation contained some elements of usefulness for Tesla because of the tremendous prestige of the financier. When, however, Tesla later found himself critically in need of funds, and it became apparent that Morgan was not financially involved in the project and apparently was not coming to the rescue of the inventor, then the reaction set in and the situation became distinctly and definitely unsatisfactory.

In 1900, however, Tesla had $150,000 on hand and a gigantic idea to be put into operation. The world-shaking superman, riding his tidal wave of fame and popularity, set to work.

# THIRD PART

## INTERNAL VIBRATION

# TWELVE

THE year 1900 marked to Tesla not only the opening of a new century but also the beginning of the world-superpower and radio-broadcasting era. With the encouragement of J. P. Morgan to spur him on—if he could accommodate any more spurring than his own inner drive furnished—and with $150,000 in cash from the same source, he was set to embark upon a gigantic venture, the building of a world wireless-power and a world broadcasting station.

The cash on hand would be totally inadequate to finance the project to completion, but this did not deter him from making a start. He needed a laboratory both to replace the Houston Street establishment, which had become entirely inadequate, and to include equipment of the type employed at Colorado Springs, but designed for use in the actual world-broadcasting process. The location was determined as the result of an arrangement he made with James S. Warden, manager and director of the Suffolk County Land Company, a lawyer and banker from the West who had acquired two thousand acres of land at Shoreham, in Suffolk County, Long Island, about sixty miles from New York. The land was made the basis of a real-estate development under the name Wardencliff.

Tesla visualized a power-and-broadcasting station which would employ thousands of persons. He undertook the establishment, eventually, of a Radio City, something far more ambitious than the enterprise in Rockefeller Center in New York which bears this name today. Tesla planned to have all wavelength channels broadcast from a single station, a project which would have given him a complete monopoly of the radio-broadcasting business. What an opportunity near-sighted businessmen of his day overlooked in not getting in on his project! But in that day Tesla was about the only one who visualized modern broadcasting. Everyone else visualized wireless as being useful only for sending telegraphic communications between ship and shore and across the ocean.

Mr. Warden saw possibilities of a sort in Tesla's plan, however, and offered him a tract of two hundred acres, of which twenty acres were cleared, for his power station, with the expectation that the two thousand men who would shortly be employed in the station would build homes on convenient sites in the remainder of the 2,000-acre tract. Tesla accepted.

Stanford White, the famous designer of many churches and other architectural monuments throughout the country, was one of Tesla's friends. He now disclosed to the famous architect his vision of an industrial "city beautiful" and sought his co-operation in realizing his dream. Mr. White was enthusiastic about the idea and, as his contribution to Tesla's work, offered to underwrite the cost of designing the strange tower the inventor sketched, and all of the architectural work involved in the general plan for the city. The actual work was done by W. D. Crow, of East Orange, N. J., one of Mr. White's associates, who later became famous as a designer of hospitals and other institutional buildings.

It was a fantastic-looking tower, with strange structural limitations, which Mr. Crow found himself designing. Tesla required a tower, about 154 feet high, to support at its peak a giant copper electrode 100 feet in diameter and shaped like a gargantuan doughnut with a tubular diameter of twenty feet. (This was later changed to a hemispherical electrode.)

The tower would have to be a skeletonized structure, built almost entirely of wood, metal to be reduced to an utter minimum and any metal fixtures employed to be of copper. No engineering data were available on wood structures of this height and type.

The structure Tesla required had a large amount of "sail area," or surface exposed to wind, concentrated at the top, creating stresses that had to be provided for in a tower that itself possessed only limited stability. Mr. Crow solved the engineering problems and then the equally difficult task of incorporating esthetic qualities in such an edifice.

When the design was completed another difficulty was encountered. None of the well-known builders could be induced to undertake the task of erecting the tower. A competent framer, associated with Norcross Bros., who were a large contracting firm in those days, finally took over the contract, although he, too, expressed fears that the winter gales might overturn the structure.* The tower was completed in 1902, and with it a large low brick building more than 100 feet square which would provide quarters for the powerhouse and laboratory. While the structures were being built, Tesla commuted every day from the Waldorf-Astoria to Wardencliff, arriving at the near-by Shoreham station shortly after eleven A.M. and remaining until three-thirty. He was always accompanied by a man servant, a Serbian, who carried a heavy hamper filled with food. When the laboratory transferred from Houston Street was in full operation at Wardencliff, Tesla rented the Bailey cottage near the Long Island Sound shore and there made his home for a year.

The heavy equipment, the dynamos and motors, that Tesla desired for his plant were of an unusual design not produced by manufacturers, and he encountered many vexatious delays in securing such material. He was able to carry on a wide range of high-frequency current and other experiments in his new laboratory, but the principal project, that of setting up the world-wide broadcasting station, lagged. Meanwhile, he had a number of glass blowers making tubes for use in transmitting and receiv-

* It stood, however, for a dozen years. When the Government, for military reasons, decided it was necessary to remove this conspicuous landmark during the First World War, heavy charges of dynamite were necessary in order to topple it, and even then it remained intact on the ground like a fallen Martian invader out of Wells' *War of the Worlds.*

ing his broadcast programs. This was a dozen years before De Forest invented the form of radio tube now in general use. The secret of Tesla's tubes died with him.

Tesla seemed to be entirely fearless of his high-frequency currents of millions of volts. He had, nevertheless, the greatest respect for the electric current in all forms, and was extremely careful in working on his apparatus. When working on circuits that might come "alive," he always worked with one hand in his pocket, using the other to manipulate tools. He insisted that all of his workers do likewise when working on the 60-cycle low-frequency alternating-current circuits, whether the potential was 50,000 or 110 volts. This safeguard reduced the possibility of a dangerous current finding a circuit through the arms across the body, where there was chance that it might stop the action of the heart.

In spite of the great care which he manifested in all of his experimental work, he had a narrow escape from losing his life at the Wardencliff plant. He was making experiments on the properties of small-diameter jets of water moving at high velocity and under very high pressures, of the order of 10,000 pounds per square inch. Such a stream could be struck by a heavy iron bar without the stream being disrupted. The impinging bar would bounce back as if it had struck another solid iron bar—a strange property for a mechanically weak substance like water. The cylinder holding the water under high pressure was a heavy one made of wrought iron. Tesla was unable to secure a wrought-iron cap for the upper surface, so he used a heavier one of cast iron, a more brittle metal. One day when he raised the pressure to a point higher than he had previously used, the cylinder exploded. The cast-iron cap broke and a large fragment shot within a few inches of his face as it went on a slanting path upward and finally crashed through the roof. The high-pressure stream of water had peculiar destructive effects on anything with which it came in contact, even tough, strong metals. Tesla never revealed the purpose or the results of these high-pressure experiments.

Tesla's insistence on the utmost neatness in his laboratory almost resulted in a tragedy through a case of thoughtlessness on the part of an assistant. Arrangements were being made for installing a heavy piece of machinery which was to be lag bolted to the thick concrete floor. Holes had been drilled in the concrete. The plan called for pouring molten lead into these holes and screwing the heavy bolts into the metal when it cooled. As soon as the holes were drilled, a young assistant starting cleaning up the débris. He not only swept up the stone chips and dust: he got a mop and thoroughly washed that area of the floor, thoughtlessly letting some of the water get into the holes. He then dried the floor. In the meantime Tesla and George Scherff, who was his financial secretary but also served in any way in which he could be helpful, were melting the lead which would hold the lag screws in the holes in the floor. Scherff took the first large ladleful of lead from the furnace and started across the laboratory to where the holes had been drilled, followed shortly by Tesla bearing another ladle.

Scherff bent down—and as he poured the hot liquid metal into one of the holes an explosion followed instantly. The molten lead was blown upward into his face in a shower of searing hot drops of liquid metal. The water which the assistant used to swab the floor had settled into the holes and, when the melted lead come in contact with it, it was changed to steam which shot the lead out of the hole like a bullet out of the barrel of a rifle. Both men were showered with drops of hot metal and dropped their ladles. Tesla, being several feet away, was only slightly injured; but Scherff was very seriously burned about the face and hands. Drops of the metal had struck his eyes and so severely burned them that it was feared for a while that his sight could not be saved.

However, despite the almost unlimited possibilities for accidents in connection with the vast variety of experiments which Tesla conducted in totally unexplored fields, using high voltages, high amperages, high pressures, high velocities and high temperatures, he went through his entire career with only one accident in which he suffered injury. In that a sharp instrument slipped, entered his palm and penetrated through the hand. The accident to Scherff was the only one in which a member of his staff was injured, with the exception of a young assistant who developed X-ray burns. He had probably been exposed to the rays from one of Tesla's tubes which, unknown to Tesla and everyone else, had been producing them even before Roentgen announced their discovery. Tesla had given them another name and had not fully investigated their properties. This was probably the first case of X-ray burns on record.

Tesla was an indefatigable worker, and it was hard for him to understand why others were incapable of such feats of endurance as he was able to accomplish. He was willing to pay unusually high wages to workers who were willing to stick with him on protracted tasks but never demanded that anyone work beyond a reasonable day's labor. On one occasion a piece of long-awaited equipment arrived and Tesla was anxious to get it installed and operating as quickly as possible. The electricians worked through twenty-four hours, stopping only for meals, and then for another twenty-four hours. The workers then dropped out, one by one, picking out nooks in the building in which to sleep. While they took from eight to twelve hours' sleep, Tesla continued to work; and when they came back to the job Tesla was still going strong and worked with them through his third sleepless twenty-four-hour period. The men were then given several days off in which to rest up; but Tesla, apparently none the worse for his seventy-two hours of toil, went through his next day of experiments, accomplishing a total of eighty-four hours without sleep or rest.

The plant at Wardencliff was intended primarily for demonstrating the radio-broadcasting phase of his "World System"; the power-distribution station was to be built at Niagara Falls.

Tesla at this time published a brochure on his "World System" which indicates the remarkable state of advancement he had projected in the wireless art, now called radio, while other experimenters were struggling to acquire familiarity with rudimentary devices. At that time, however, his promises seemed fantastic. The brochure contained the following description of his system and his objectives:

The *World System* has resulted from a combination of several original discoveries made by the inventor in the course of long continued research and experimentation. It makes possible not only the instantaneous and precise wireless transmission of any kind of signals, messages or characters, to all parts of the world, but also the interconnection of the existing telegraph, telephone, and other signal stations without any change in their present equipment. By its means, for instance, a telephone subscriber here many call up any other subscriber on the Globe. An inexpensive receiver, not bigger than a watch, will enable him to listen anywhere, on land or sea, to a speech delivered, or music played in some other place, however distant. These examples are cited merely to give an idea of the possibilities of this great scientific advance, which annihilates distance and makes that perfect conductor, the Earth, available for all the innumerable purposes which human ingenuity has found for a line wire. One far reaching result of this is that any device capable of being operated through one or more wires (at a distance obviously restricted) can likewise be actuated, without artificial conductors and with the same facility and accuracy, at distances to which there are no limits other than those imposed by the physical dimensions of the Globe. Thus, not only will entirely new fields for commercial exploitation be opened up, by this ideal method of transmission, but the old ones vastly extended.

The World System is based on the application of the following important inventions and discoveries:

1. The *Tesla Transformer*. This apparatus is, in the production of electrical vibrations, as revolutionary as gunpowder was in warfare. Currents many times stronger than any ever generated in the usual ways, and sparks over 100 feet long have been produced by the inventor with an instrument of this kind.

2. The *Magnifying Transmitter*. This is Tesla's best invention—a peculiar transformer specially adapted to excite the Earth, which

is in the transmission of electrical energy what the telescope is in astronomical observation. By the use of this marvelous device he has already set up electrical movements of greater intensity than those of lightning and passed a current, sufficient to light more than 200 incandescent lamps, around the Globe.

3. The *Tesla Wireless System*. This system comprises a number of improvements and is the only means known for transmitting economically electrical energy to a distance without wires. Careful tests and measurements in connection with an experimental station of great activity, erected by the inventor in Colorado, have demonstrated that power in any desired amount can be conveyed clear across the Globe if necessary, with a loss not exceeding a few per cent.

4. The *Art of Individualization*. This invention of Tesla is to primitive tuning what refined language is to unarticulated expression. It makes possible the transmission of signals or messages absolutely secret and exclusive both in active and passive aspect, that is, non-interfering as well as non-interferable. Each signal is like an individual of unmistakable identity and there is virtually no limit to the number of stations or instruments that can be simultaneously operated without the slightest mutual disturbance.

5. The *Terrestrial Stationary Waves*. This wonderful discovery, popularly explained, means that the Earth is responsive to electrical vibrations of definite pitch just as a tuning fork to certain waves of sound. These particular electrical vibrations, capable of powerfully exciting the Globe, lend themselves to innumerable uses of great importance commercially and in many other respects.

The first World System power plant can be put in operation in nine months. With this power plant it will be practical to attain electrical activities up to ten million horsepower and it is designed to serve for as many technical achievements as are possible without undue expense. Among these the following may be mentioned:

1. Interconnection of the existing telegraph exchanges of offices all over the World;

2. Establishment of a secret and non-interferable government telegraph service;

3. Interconnection of all the present telephone exchanges or offices all over the Globe;

4. Universal distribution of general news, by telegraph or telephone, in connection with the Press;

5. Establishment of a World System of intelligence transmission for exclusive private use;

6. Interconnection and operation of all stock tickers of the world;

7. Establishment of a world system of musical distribution, etc.;

8. Universal registration of time by cheap clocks indicating the time with astronomical precision and requiring no attention whatever;

9. Facsimile transmission of typed or handwritten characters, letters, checks, etc.;

10. Establishment of a universal marine service enabling navigators of all ships to steer perfectly without compass, to determine the exact location, hour and speed, to prevent collisions and disasters, etc.;

11. Inauguration of a system of world printing on land and sea;

12. Reproduction anywhere in the world of photographic pictures and all kinds of drawings or records.

Thus, more than forty years ago, Tesla planned to inaugurate every feature of modern radio, and several facilities which have not yet been developed. He was to continue, for another twenty years, to be the only "wireless" inventor who had yet visualized a broadcasting service.

While at work on his Wardencliff radio-broadcasting plant, Tesla was also evolving plans for establishing his world power station at Niagara Falls. So sure was he of the successful outcome of his efforts that he stated in a newspaper interview in 1903 that he would light the lamps of the coming international exposition in Paris with power wirelessly transmitted from the Falls. Circumstances, however, prevented him from making good this promise. His difficulties and his plans were outlined in a statement published in the *Electrical World and Engineer*, March 5, 1904:

The first of these central plants would have been already completed had it not been for unforeseen delays which, fortunately, have nothing to do with its purely technical features. But this loss of time, while vexatious, may, after all, prove to be a blessing in disguise. The best design of which I know has been adopted, and the transmitter will emit a wave complex of a total maximum activity of 10,000,000 horsepower, one percent of which is amply sufficient to "girdle the globe." This enormous rate of energy delivery, approximately twice that of the combined falls of Niagara, is obtainable only by the use of certain artifices, which I shall make known in due course.

For a large part of the work which I have done so far I am indebted to the noble generosity of Mr. J. Pierpont Morgan, which was all the more welcome and stimulating, as it was extended at a time when those, who have since promised most, were the greatest of doubters. I have also to thank my friend Stanford White, for much unselfish and valuable assistance. This work is now far advanced, and though the results may be tardy, they are sure to come.

Meanwhile, the transmission of energy on an industrial scale is not being neglected. The Canadian Niagara Power Company have offered me a splendid inducement, and next to achieving success for the sake of the art, it will give me the greatest satisfaction to make their concession financially profitable to them. In this first power plant, which I have been designing for a long time, I propose to distribute 10,000 horsepower under a tension of 10,000,000 volts, which I am now able to produce and handle with safety.

This energy will be collected all over the globe preferably in small amounts, ranging from a fraction of one to a few horsepower. One of the chief uses will be the illumination of isolated homes. It takes very little power to light a dwelling with vacuum tubes operated by high frequency currents and in each instance a terminal a little above the roof will be sufficient. Another valuable application will be the driving of clocks and other such apparatus. These clocks will be exceedingly simple, will require absolutely no attention and will indicate rigorously correct time. The idea of impressing upon the earth American time is fascinating and very likely to become popular. There are innumerable devices of all kinds which are either now employed or can be supplied and by operating them in this manner I may be able to offer a great convenience to the whole world with a plant of no more than 10,000 horsepower. The introduction of this system will give opportunities for invention and manufacture such as have never presented themselves before.

Knowing the far reaching importance of this first attempt and its effect upon future development, I shall proceed slowly and carefully. Experience has taught me not to assign a term to enterprises the consummation of which is not wholly dependent on my own abilities and exertions. But I am hopeful that these great realizations are not far off and I know that when this first work is completed they will follow with mathematical certitude.

When the great truth accidentally revealed and experimentally confirmed is fully recognized, that this planet, with all its appalling immensity, is to electric current virtually no more than a small metal ball and that by this fact many possibilities, each baffling the imagination and of incalculable consequence, are rendered absolutely sure of accomplishment; when the first plant is inaugurated, and it is shown that a telegraphic message, almost as secret and non-interferable as a thought, can be transmitted to any terrestrial distance, the sound of the human voice, with all its intonations and inflections, faithfully and instantly reproduced at any point of the globe, the energy of a waterfall made available for supplying light, heat or motive power, anywhere—on sea, or land, or high in the air—humanity will be like an ant heap stirred up with a stick: See the excitement coming.

The Niagara Falls plant was never built; and difficulties, soon enough, were encountered at the Wardencliff plant not only in securing desired equipment but also finances.

Tesla's greatest oversight was that he neglected to invent, so to speak, a device for making the unlimited quantities of money that were necessary to develop his other inventions. As we have seen, he was utterly lacking in the phase of personality that made possible the securing of financial returns directly from his inventions. An individual with his ability could have made millions out of each of a number of Tesla's minor inventions. If he had taken the trouble, for example, to collect annual royalties on twenty or more different kinds of devices put out by as many manufacturers employing his Tesla coil for medical treatments, he would have had ample income to finance his World Wireless System.

His mind, however, was too fully occupied with fascinating scientific problems. He had, at times, nearly a score of highly skilled workmen constantly employed in his laboratory developing the electrical inventions he was continuing to make at a rapid rate. Armed guards were always stationed around the laboratory to prevent spying on his inventions. His payroll was heavy, his bank balance became dangerously low, but he was so immersed in his experimental work that he continuously put off the task of making an effort to repair his finances. He soon found himself facing judgments obtained by creditors on accounts upon which he could not make payments. He was forced, in 1905, to close the Wardencliff laboratory.

The fantastic tower in front of the laboratory was never completed. The doughnut-shaped copper electrode was never built because Tesla changed his mind and decided to have a copper hemisphere 100 feet in diameter and 50 feet high built on top of the 154-foot cone-shaped tower. A skeleton framework for holding the hemispherical plates was built, but the copper sheeting was never applied to it. The 300-horsepower dynamos and the apparatus for operating the broadcasting station were left intact, but they were eventually removed by the engineering firm that installed them and had not been paid.

Tesla opened an office at 165 Broadway, in New York, where for a while he tried to contrive some means for reviving his project. Thomas Fortune Ryan, the well-known financier, and H. O. Havemeyer, the leading sugar refiner, aided him with contributions of $10,000 and $5,000 respectively. Instead of using these to open another laboratory, he applied them to paying off the debts on his now defunct World Wireless System. He paid off every penny due to every creditor.

When it became apparent that Tesla was in financial difficulties, many who had assumed that Morgan was financially involved as an investor in his project were disillusioned. When specific inquiries revealed that the great financier held no interest whatever in the enterprise, the rumor got into circulation that Morgan had withdrawn his support; and when no reason for such action could be learned the rumor expanded to carry the story that Tesla's system was impracticable. As a matter of fact, Morgan continued to make generous personal contributions to Tesla almost up to the time of his own death; and his son did so to a lesser extent for a short time.

Tesla made no effort to combat the growing rumors.

If Tesla could have tolerated a business manager, and had

placed the development of his patents in the hands of a businessman, he could have established as early as 1896 a practical ship-to-
shore, and probably a trans-oceanic wireless service; and these
would have given him a monopoly in this field. He was asked
to rig up a wireless set on a boat to report the progress of the
international yacht race for Lloyds of London in 1896, but he
refused the offer, which was a lucrative one, on the grounds that
he would not demonstrate his system publicly on less than a
world-wide basis because it could be confused with the amateur-
ish efforts being made by other experimenters. If he had accepted
this offer—and he could have met the requirements without the
least technical difficulty—he undoubtedly would have found his
interests diverted to some extent into a profitable commercial
channel that might have made a vast, and favorable, change in
the second half of his life.

Tesla, however, could not be bothered with minor, even
though profitable, projects. The superman, the man magnificent,
was too strong in him. The man who had put industry on an
electrical power basis, the man who had set the whole earth in
vibration, could not fill a minor rôle of carrying messages for
hire. He would function in his major capacity or not at all; he
would be a Jupiter, never a Mercury.

George Scherff, who was engaged by Tesla as bookkeeper and
secretary when he opened his Houston Street laboratory, was a
practical individual. He managed, as far as was humanly possible, to keep the inventor disentangled in his contacts with the
business world. The more he knew Tesla, the better he liked
him; and the more respect he had for his genius and his ability
as an inventor, the more he became conscious of the fact that
this genius was totally lacking in business ability.

Scherff was understandably distressed by a situation in which
an enterprise was continuously spending money but never receiving any. He sought to protect as far as possible the $40,000
which Tesla received from Adams as an investment in the enterprise; and it was stretched to cover more than three years of great

*215*

activity. Scherff wanted Tesla to work out plans for deriving an income from his inventions. Each new development which Tesla produced was studied by Scherff and made the basis for a plan for manufacture and sale of a device. Tesla uniformly rejected all the suggestions. "This is all small-time stuff," he would reply. "I cannot be bothered with it."

Even when it was pointed out to him that many manufacturers were using his Tesla coils, selling great numbers of them and making plenty of money out of them, his interest could not be aroused to enter this profitable field, nor to permit Scherff to arrange to have a sideline set-up which could be conducted without interfering with his research work. Nor could he be induced to bring suits to protect his invention and seek to make the manufacturers pay him royalties. He admitted, however, "If the manufacturers paid me twenty-five cents on each coil they sold I would be a wealthy man."

When Lloyds of London made their request that he set up a wireless outfit on a boat and report the international yacht races of 1896, by his new wireless system, and offered a generous honorarium, Scherff became insistent that the offer be accepted; and he urged Tesla to drop all other work temporarily and use the publicity he would get from the exploit as a means of floating a commercial company for transmitting wireless messages between ship and shore and across the ocean, pointing out that money would be made both in manufacturing the apparatus and in transmitting messages. The company, Scherff suggested, could be operated by managers to produce an income and Tesla could return to his work of making inventions and always have plenty of money to pay for the cost of his researches.

Scherff can look back today, as he sits on the porch of his Westchester home, and decide, through a retrospect of fifty years, that his plan was basically sound, with the Radio Corporation of America, its extensive manufacturing facilities and its worldwide communication system, its tremendous capital system and earnings, as evidence in support of the claim.

*216*

Tesla's reply to the proposal was, as usual, "Mr. Scherff, that is small-time stuff. I cannot be bothered with it. Just wait until you see the magnificent inventions I am going to produce, and then we will all make millions."

Tesla's millions never came. Scherff remained with him until the Wardencliff laboratory closed, owing to the lack of income, which he had been trying to circumvent. Scherff then established a lucrative connection with the Union Sulphur Co. but he still continued, without taking compensation, to give Tesla one day a week of his time and keep his business affairs disentangled as far as possible. Tesla was meticulously careful about paying everyone who performed any service for him, but this was counterbalanced by an active faculty for contracting bills without waiting to see if he had funds on hand to meet them. Money was an annoying anchor that always seemed to be dragging and hindering his research activities—something that was too mundane to merit the time and attention he should be giving to more important things.

Scherff, tight-lipped and businesslike, cannot be induced to talk of Tesla's affairs. If he were, instead, a loquacious philosopher, he might be induced to smile over the frailties of human nature, and the strange pranks which fate can play on individuals, as he thinks of Tesla, who, on the basis of a single invention, might have become an individual Radio Corporation of America and failed to do so, and who passed up equal chances on two hundred other inventions, any one of which could have produced a fortune. And for contrast, he can recall occasions in recent decades when it was necessary to make modest loans to the great Tesla to permit him to meet the need for current personal necessities. But Scherff refuses to permit any close questions or discussion about these incidents.

# THIRTEEN

WHEN his World Wireless System project crashed, Tesla turned again to a project to which he had given considerable thought at the time he was developing his polyphase alternating-current system: that of developing a rotary engine which would be as far in advance of existing steam engines as his alternating-current system was ahead of the direct-current system, and which could be used for driving his dynamos.

All of the steam engines in use in powerhouses at that time were of the reciprocating type; essentially the same as those developed by Newcomer and Watt, but larger in size, better in construction and more efficient in operation.

Tesla's engine was of a different type—a turbine in which jets of steam injected between a series of disks produced rotary motion at high velocity in the cylinder on which these disks were mounted. The steam entered at the outer edge of the disks, pursued a spiral path of a dozen or more convolutions, and left the engine near the central shaft.

When Tesla informed a friend in 1902 that he was working on an engine project, he declared he would produce an engine so small, simple and powerful that it would be a "powerhouse in a hat." The first model, which he made about 1906, fulfilled this promise. It was small enough to fit into the dome of a derby hat, measured a little more than six inches in its largest dimension, and developed thirty horsepower. The power-producing performance of this little engine vastly exceeded that of every known kind of prime mover in use at that time. The engine weighed a little less than ten pounds. Its output was therefore three horsepower per pound. The rotor weighed only a pound and a half, and its light weight and high power yield gave Tesla a slogan which he used on his letterheads and envelopes—"Twenty horsepower per pound."

There was nothing new, of course, in the basic idea of obtaining circular motion directly from a stream of moving fluid. Windmills and water wheels, devices as old as history, performed this feat. Hero, the Alexandrian writer, about 200 B.C., described, if he did not invent, the first turbine. It consisted of a hollow sphere of metal mounted on an axle, with two tubes sticking out of the sphere at a tangent to its surface. When water was placed in the sphere and the device was suspended in a fire, the reaction of the steam coming out of the tubes caused the device to rotate.

Tesla's ingenious and original development of the turbine idea probably had its origin in that amusing and unsuccessful experiment he made when, as a boy, he tried to build a vacuum motor and observed its wooden cylinder turn slightly by the drag of the air leaking into the vacuum chamber. Later, too, when as a youth he fled to the mountains to escape military service and played with the idea of transporting mail across the ocean through an underwater tube, through which a hollow sphere was to be carried by a rapidly moving stream of water, he had discovered that the friction of the water on the walls of the tube made the idea impracticable. The friction would slow down the velocity of the stream of water so that excessive amounts of power would be required to move the water at a desired speed and pressure. Conversely, if the water moved at this speed, the friction caused it to try to drag the enclosing tube along with it.

It was this friction which Tesla now utilized in his turbine. A jet of steam rushing at high velocity between disks with a very small distance separating them was slowed down by the friction —but the disks, being capable of rotation, moved with increasing velocity until it was almost equal to that of the steam. In addition to the friction factor, there exists a peculiar attraction between gases and metal surfaces; and this made it possible for the moving steam to grip the metal of the disks more effectively and drag them around at high velocities. The first model which Tesla made in 1906 had twelve disks five inches in diameter. It was operated by compressed air, instead of steam, and attained a speed of 20,000 revolutions per minute. It was Tesla's intention eventually to use oil as fuel, burning it in a nozzle and taking advantage of the tremendous increase in volume, in the change from a liquid to burned highly expanded gases, to turn the rotor. This would eliminate the use of boilers for generating steam and give the direct process proportional increased efficiency.

Had Tesla proceeded with the development of his turbine in 1889 when he returned from the Westinghouse plant, his turbine might perhaps have been the one eventually developed to replace the slow, big, lumbering reciprocating engines then in use. The fifteen years, however, which he devoted to the development of currents of high potential and high frequency, had entailed a delay which gave opportunity for developers of other turbine ideas to advance their work to a stage which now was effective in putting Tesla in the status of a very late starter. In the meantime, turbines had been developed which were virtually windmills in a box. They consisted of rotors with small buckets or vanes around the circumference which were struck by the incoming steam jet. They lacked the simplicity of the Tesla turbine; but by the time Tesla introduced his type, the others were well entrenched in the development stage.

Tesla's first tiny motor was built in 1906 by Julius C. Czito, who operated at Astoria, Long Island, a machine shop for making inventor's models. He also built the subsequent 1911 and 1925 models of the turbine, and many other devices on which Tesla worked up to 1929. Mr. Czito's father had been a member of Tesla's staff in the Houston Street laboratories, from 1892 to 1899, and at Colorado Springs.

Mr. Czito's description of the first model is as follows:

The rotor consisted of a stack of very thin disks six inches in diameter, made of German silver. The disks were one thirty-second of an inch thick and were separated by spacers of the same metal and same thickness but of much smaller diameter which were cut in the form of a cross with a circular center section. The extended arms served as ribs to brace the disks.

There were eight disks and the edgewise face of the stack was only one-half inch across. They were mounted on the center of a shaft about six inches long. The shaft was nearly an inch in diameter in the mid section and was tapered in steps to less than half an inch at the ends. The rotor was set in a casing made in four parts bolted together.

The circular chamber where the rotor turned was accurately machined to allow a clearance of one sixty-fourth of an inch between the casing and the face of the rotor. Mr. Tesla desired an almost touching fit between the rotor face and the casing when the latter was turning. The large clearance was necessary because the rotor attained tremendously high speeds, averaging 35,000 revolutions per minute. At this speed the centrifugal force generated by the turning movement was so great it appreciably stretched the metal in the rotating disks. Their diameter when turning at top speed was one thirty-second of an inch greater than when they were standing still.

A larger model was built by Tesla in 1910. It had disks twelve inches in diameter, and with a speed of 10,000 revolutions per minute it developed 100 horsepower, indicating a greatly improved efficiency over the first model. It developed more than three times as much power at half the speed.

During the following year, 1911, still further improvements were made. The disks were reduced to a diameter of 9.75 inches and the speed of operation was cut down by ten per cent, to 9,000 revolutions per minute—and the power output increased by ten per cent, to 110 horsepower!

Following this test, Tesla issued a statement in which he declared:

I have developed 110 horsepower with disks nine and three quarter inches in diameter and making a thickness of about two inches. Under proper conditions the performance might have been as much as 1,000 horsepower. In fact there is almost no limit to the mechanical performance of such a machine. This engine will work with gas, as in the usual type of explosion engine used in automobiles and airplanes, even better than it did with steam. Tests which I have conducted have shown that the rotary effort with gas is greater than with steam.

Enthusiastic over the success of his smaller models of the turbine, operated on compressed air, and to a more limited extent by direct combustion of gasoline, Tesla designed and built a larger, double unit, which he planned to test with steam in the Waterside Station, the main powerhouse of the New York Edison Company.

This was a station which had originally been designed to operate on the direct-current system developed by Edison—but it was now operating throughout on Tesla's polyphase alternating-current system.

Now Tesla, invading the Edison sanctum to test a new type of turbine which he hoped would replace the types in use, was definitely in enemy territory. The fact that he had Morgan backing, and that the Edison Company was a "Morgan company," had no nullifying effect on the Edison-Tesla feud.

This situation was not softened in any way by Tesla's method of carrying on his tests. Tesla was a confirmed "sun dodger"; he preferred to work at night rather than in the daytime. Powerhouses, not from choice but from necessity, have their heaviest demands for current after sunset. The day load would be relatively light; but as darkness approached, the dynamos started to groan under the increasing night load. The services of the workers at the Waterside Station were made available to Tesla for the setting up and tests of his turbine with the expectation that the work would be done during the day when the tasks of the workers were easiest.

Tesla, however, would rarely show up until five o'clock in the afternoon, or later, and would turn a deaf ear to the pleas of workers that he arrive earlier. He insisted that certain of the workers whom he favored remain after their five-o'clock quitting time on the day shift to work with him on an overtime basis. Nor did he maintain a conciliatory attitude toward the engineering staff or the officials of the company. The attitudes, naturally, were mutual.

The turbine Tesla built for this test had a rotor 18 inches in diameter which turned at a speed of 9,000 revolutions per minute. It developed 200 horsepower. The overall dimensions of the engine were—three feet long, two feet wide and two feet high. It weighed 400 pounds.

Two such turbines were built and installed in a line on a single base. The shafts of both were connected to a torque rod. Steam was fed to both engines so that, if they were free to rotate, they would turn in opposite directions. The power developed was measured by the torque rod connected to the two opposing shafts.

At a formal test, to which Tesla invited a great many guests, he issued a statement in which he said, as reported, in part:

It should be noted that although the experimental plant develops 200 horsepower with 125 pounds at the supply pipe and free exhaust it could show an output of 300 horsepower with full pressure of the supply circuit. If the turbine were compounded and the exhaust were led to a low pressure unit carrying about three times the number of disks contained in the high pressure element, with connection to a condenser affording 28.5 to 29.0 inches of vacuum the results obtained in the present high pressure machine indicate that the compounded unit would give an output of 600 horsepower without great increase of dimensions. This estimate is very conservative.

Tests have shown that when the turbine is running at 9,000 revolutions per minute under an inlet pressure of 125 pounds to the square inch and with free exhaust 200 brake horsepower are developed. The consumption under these conditions of maximum output is 38 pounds of saturated steam per horsepower per hour, a very high efficiency when we consider that the heat drop, measured by thermometers, is only 130 B.T.U. and that the energy transformation is effected in one stage. Since three times the number of heat units are available in a modern plant with superheat and high vacuum the utilization of these facilities would mean a consumption of less than 12 pounds per horsepower hour in such turbines adapted to take the full drop.

Under certain conditions very high thermal efficiencies have been obtained which demonstrate that in large machines based on this principle steam consumption will be much lower and should approximate the theoretical minimum thus resulting in the nearly frictionless turbine transmitting almost the entire expansive energy of the steam to the shaft.

It should be kept in mind that all of the turbines which Tesla built and tested were single-stage engines, using about one-third of the energy of the steam. In practical use, they were intended to be installed with a second stage which would employ the remaining energy and increase the power output about two or three fold. (The two types of turbines in common use each have a dozen and more stages within a single shell.)

Some of the Edison electric camp, observing the torque-rod tests and apparently not understanding that in such a test the two rotors remain stationary—their opposed pressures staging a tug of war measured as torque—circulated the story that the turbine was a complete failure; that this turbine would not be practical if its efficiency had been increased a thousand fold. It was stories such as these that contributed to the imputation that Tesla was an impractical visionary. The Tesla turbine, however, used as a single-stage engine, functioning as a pygmy power producer, in the form in which it was actually tested, anticipated by more than twenty-five years a type of turbine which has been installed in recent years in the Waterside Station. This is a very small engine, with blades on its rotor, known as a "topping turbine," which is inserted in the steam line between the boilers and the ordinary turbines. Steam of increased pressure is supplied, and the topping turbine skims this "cream" from the steam and exhausts steam that runs the other turbines in their normal way.

The General Electric Company was developing the Curtis turbine at that time, and the Westinghouse Electric and Manufacturing Company was developing the Parsons turbine; and neither company showed the slightest interest in Tesla's demonstration.

Further development of his turbine on a larger scale would

have required a large amount of money—and Tesla did not possess even a small amount.

FINALLY he succeeded in interesting the Allis Chalmers Manufacturing Company of Milwaukee, builders of reciprocating engines and turbines, and other heavy machinery. In typical Tesla fashion, though, he manifested in his negotiations such a lack of diplomacy and insight into human nature that he would have been better off if he had completely failed to make any arrangements for exploiting the turbine.

Tesla, an engineer, ignored the engineers on the Allis Chalmers staff and went directly to the president. While an engineering report was being prepared on his proposal, he went to the Board of Directors and "sold" that body on his project before the engineers had a chance to be heard. Three turbines were built. Two of them had twenty disks eighteen inches in diameter and were tested with steam at eighty pounds pressure. They developed at speeds of 12,000 and 10,000 revolutions per minute, respectively, 200 horsepower. This was exactly the same power output as had been achieved by Tesla's 1911 model, which had disks of half this diameter and was operated at 9,000 revolutions under 125 pounds pressure. A much larger engine was tackled next. It had fifteen disks sixty inches in diameter, was designed to operate at 3,600 revolutions per minute, and was rated at 500 kilowatts capacity, or about 675 horsepower.

Hans Dahlstrand, Consulting Engineer of the Steam Turbine Department, reports, in part:

We also built a 500 kw steam turbine to operate at 3,600 revolutions. The turbine rotor consisted of fifteen disks 60 inches in diameter and one eighth inch thick. The disks were placed approximately one eighth inch apart. The unit was tested by connecting to a generator. The maximum mechanical efficiency obtained on this unit was approximately 38 per cent when operating at steam pressure of approximately 80 pounds absolute and a back pressure of approximately 3 pounds absolute and 100 degrees F superheat at the inlet.

When the steam pressure was increased above that given the

mechanical efficiency dropped, consequently the design of these turbines was of such a nature that in order to obtain maximum efficiency at high pressure, it would have been necessary to have more than one turbine in series.

The efficiency of the small turbine units compares with the efficiency obtainable on small impulse turbines running at speeds where they can be directly connected to pumps and other machinery. It is obvious, therefore, that the small unit in order to obtain the same efficiency had to operate at from 10,000 to 12,000 revolutions and it would have been necessary to provide reduction gears between the steam turbine and the driven unit.

Furthermore, the design of the Tesla turbine could not compete as far as manufacturing costs with the smaller type of impulse units. It is also questionable whether the rotor disks, because of light construction and high stress, would have lasted any length of time if operating continuously.

The above remarks apply equally to the large turbine running at 3,600 revolutions. It was found when this unit was dismantled that the disks had distorted to a great extent and the opinion was that these disks would ultimately have failed if the unit had been operated for any length of time.

The gas turbine was never constructed for the reason that the company was unable to obtain sufficient engineering information from Mr. Tesla indicating even an approximate design that he had in mind.

Tesla appears to have walked out on the tests at this stage. In Milwaukee, however, there was no George Westinghouse to save the situation. Later, during the twenties, the author asked Tesla why he had terminated his work with the Allis Chalmers Company. He replied: "They would not build the turbines as I wished"; and he would not amplify the statement further.

The Allis Chalmers Company later became the pioneer manufacturers of another type of gas turbine that has been in successful operation for years.

While the Dahlstrand report may appear to be severely critical of the Tesla turbine and to reveal fundamental weaknesses in it not found in other turbines, such is not the case. The report is, in general, a fair presentation of the results; and the descrip-

226

tion of apparent weaknesses merely offers from another viewpoint the facts which Tesla himself stated about the turbine in his earlier test—that when employed as a single-stage engine it uses only about a third of the energy of the steam, and that to utilize the remainder, it would have to be compounded with a second turbine.

The reference to a centrifugal force of 70,000 pounds resulting from the high speed of rotation of the rotor, causing damage to the disks, refers to a common experience with all types of turbines. This is made clear in a booklet on "The Story of the Turbine," issued during the past year by the General Electric Company, in which it is stated:

It [the turbine] had to wait until engineers and scientists could develop materials to withstand these pressures and speeds. For example, a single bucket in a modern turbine travelling at 600 miles per hour has a centrifugal force of 90,000 pounds trying to pull it from its attachment on the bucket wheel and shaft. . . .

In this raging inferno the high pressure buckets at one end of the turbine run red hot while a few feet away the large buckets in the last stages run at 600 miles per hour through a storm of tepid rain— so fast that the drops of condensed steam cut like a sand blast.

Dahlstrand reported that difficulties were encountered in the Tesla turbine from vibration, making it necessary to re-enforce the disks. That this difficulty is common to all turbines is further indicated by the General Electric booklet, which states:

Vibration cracked buckets and wheels and wrecked turbines, sometimes within a few hours and sometimes after years of operation. This vibration was caused by taking such terrific amounts of power from relatively light machinery—it some cases as much as 400 horsepower out of a bucket weighing but a pound or two. . . .

The major problems of the turbine are four—high temperatures, high pressures, high speeds and internal vibration. And their solution lies in engineering, research and manufacturing skill.

These problems are still awaiting their final solution, even with the manufacturers who have been building turbines for forty years; and the fact that they were encountered in the Tesla turbine, and so reported, is not a final criticism of Tesla's invention in the earliest stages of its development.

There have been whisperings in engineering circles during the past year or two to indicate a revival of interest in the Tesla turbine and the possibility that the makers of the Curtis and Parsons types may extend their lines to include the Tesla type for joint operation with the others. The development of new alloys, which can now almost be made to order with desired qualities of mechanical stability under conditions of high temperature and great stresses, is largely responsible for this turn of events.

It is a possibility that if the Tesla turbine were constructed with the benefit of two or more stages, thus giving it the full operating range of either the Curtis or the Parsons turbine, and were built with the same benefits of engineering skill and modern metallurgical developments as have been lavished on these two turbines, the vastly greater simplicity of the Tesla turbine would enable it to manifest greater efficiencies of operation and economies of construction.

# FOURTEEN

THE highest honor which the world can confer upon its scholars is the Nobel Prize founded by Alfred B. Nobel, the Swedish scientist who gained his wealth through the invention of dynamite. Five awards are made annually, and each carries an honorarium of about $40,000, in normal times.

An announcement came from Sweden, in 1912, that Nikola Tesla and Thomas A. Edison had been chosen to share the 1912 award in physics. The awards, however, were never made; and the prize went instead to Gustav Dalen, a Swedish scientist.

The full story of what took place is not known. The correspondence on the subject is not available. It is definitely established that Tesla refused to accept the award. Tesla was very much in need of money at this time and the $20,000, which would have been his share of the divided award, would have aided him to continue his work. Other factors, however, had a more potent influence.

Tesla made a very definite distinction between the inventor of useful appliances and the discoverer of new principles. The discoverer of new principles, he stated in conversation with the author, is a pioneer who opens up new fields of knowledge into which thousands of inventors flock to make commercial applications of the newly revealed information. Tesla declared himself a discoverer and Edison an inventor; and he held the view that placing the two in the same category would completely destroy all sense of the relative value of the two accomplishments.

It is quite probable that Tesla was also influenced by the fact that the Nobel Prize in physics had been awarded to Marconi three years earlier, a situation that greatly disappointed him. To have the award go first to Marconi, and then to be asked to share the award with Edison, was too great a derogation of the relative value of his work to the world for Tesla to bear without rebelling.

Tesla was the first, and probably the only, scientist to refuse this famous prize.

One of the highest honors in the engineering world, too, is the Edison Medal, founded by unnamed friends of Thomas A. Edison, and awarded each year by the American Institute of Electrical Engineers, at its annual convention, for outstanding contribution to electrical art and science. Usually, the recipients are very happy to receive the award; but in 1917, when the committee voted to present the medal to Tesla, a different situation developed.

The chairman of the Edison Medal committee was B. A. Behrend, who had been one of the first electrical engineers to grasp the tremendous significance of Tesla's alternating-current discoveries and their far-reaching importance to every department of the electrical industry. A few outstanding engineers were able, at the beginning, to understand the intricacies of new alternating current procedures which Tesla's discoveries made of immediate practical importance; but it was Behrend who developed a beautiful, simple mathematical technique, known as the "circle diagram," which made it possible to work out problems of designing alternating-current machinery with great ease, and also to understand the complex phenomena that were taking place within such devices. He published innumerable articles on the subject in the technical journals and wrote the standard textbook on the subject, *The Induction Motor*. Fame and fortune came to Behrend. He achieved recognition as one of the outstanding electrical engineers, and was later elected vice-president of the American Institute of Electrical Engineers. So important was his work to the commercial world that he was considered a probable recipient of the Edison Medal.

Behrend had started publishing articles on his circle diagram discovery in 1896 but he did not meet Tesla until 1901, when Tesla required a particular type of motor for his World Wireless plant being built at Wardencliff, L. I., and the task of designing it was assigned to the engineering department of a manufacturing company of which Behrend was in charge. After Tesla and Behrend met, a very close personal friendship developed between the two men. Behrend was one of the few who thoroughly understood Tesla's work; and the inventor, lonely in the absence of individuals with minds of his own caliber, greatly appreciated Behrend's friendship.

Behrend believed, therefore, that he was rendering Tesla a token of his highest appreciation when he managed to maneuver the award of the Edison Medal to him; and he was quite happy to carry out the mission of bearing the good news to the inventor. The announcement, however, did not make Tesla happy. He did not want the Edison Medal, he would not receive it!

Behrend, greatly surprised at Tesla's rebuff, asked him if he would not explain the situation that caused it.

"Let us forget the whole matter, Mr. Behrend. I appreciate your good will and your friendship but I desire you to return to the committee and request it to make another selection for a recipient. It is nearly thirty years since I announced my rotating magnetic field and alternating-current system before the Institute. I do not need its honors and someone else may find them useful."

It would have been impossible for Behrend to deny that the Institute had indeed failed, over this long period, to honor the man whose discoveries were responsible for creating the jobs held by probably more than three quarters of the members of the Institute, while honors had been distributed to many others for relatively minor accomplishments. Still, using the privilege of friendship, Behrend pressed for a further explanation.

"You propose," Tesla replied, "to honor me with a medal which I could pin upon my coat and strut for a vain hour before the members and guests of your Institute. You would bestow an outward semblance of honoring me but you would decorate my body and continue to let starve, for failure to supply recognition, my mind and its creative products which have supplied the foundation upon which the major portion of your Institute exists. And when you would go through the vacuous pantomine of honoring Tesla you would not be honoring Tesla but Edison who has previously shared unearned glory from every previous recipient of this medal."

Behrend, however, after several visits, finally prevailed upon Tesla to accept the medal.

Custom requires that the recipient of a medal deliver a formal address. On the occasions, a quarter of a century earlier, when Tesla was invited to address the Institute, he had had ample laboratory facilities, and had invested a great deal of time, effort, thought and money in the preparation of his lectures. For them, however, he was awarded no honors. Now he was without laboratory facilities and without adequate financial resources, although his more mature mind was as filled with ideas and unborn inventions as it had ever been. He was not required to present a demonstration lecture. In this matter, however, Tesla was a victim of his own past performances; and there was an expectancy that he would emerge from the comparative oblivion which had enshrouded him for more than a decade, and come, like a master magician, bearing some wondrous new gifts of invention to the world.

Tesla attended some of the meetings of the convention, and Behrend, none too certain about what the medalist might do, took him in tow following the afternoon session and escorted him to the Hotel St. Regis, where Tesla now made his home, and where both donned their formal dress for the evening's ceremonies.

The first event on the evening's program was a private dinner at the Engineers' Club, tendered by the Institute to the medalist, who was the guest of honor, and attended by previous recipients of the Edison Medal, as well as members of the committee and the officers of the Institute. It was a gala occasion and represented an unusual concentration of the world's greatest electrical engineering talent. Tesla could be relied upon to lend brilliance to any such occasion, but, while his sparkling conversation added to the gayety of the group, he was distinctly ill at ease.

The Engineers' Club, on the south side of 40th Street, between Fifth and Sixth Avenues, faces Bryant Park, the eastern third of which is occupied by the classical building of the New York Public Library, facing Fifth Avenue from 40th to 42nd Streets. The United Engineering Societies Building, an imposing structure on the north side of 39th, stands almost back-to-back with the Engineers' Club. By stepping a few feet across an alley, it is possible to go from one building to the other.

Following the dinner in the Engineers' Club, the brilliant group at the medalist's dinner made their way across the alley and proceeded through the crowded lobby of the Engineering Societies Building, which was abuzz with the multitudinous activities associated with a convention. The party entered the elevators which carried them to the large auditorium on the fifth floor where the medal presentations were to take place.

The auditorium was crowded with an audience that had come largely from formal dinners held as part of the convention program. The floor and gallery were filled to capacity. The buzz of animated conversation died down as there filed onto the stage the outstanding figures of the electrical world, in "tails" and white ties, who were to serve as the "wax works" of the ceremonies and to take some part in the presentation.

As the wax works took their previously assigned chairs, the stage was set for the opening of the ceremonies. But the opening did not take place according to schedule. There was consternation in the group as it was discovered that the chair reserved for the chief participant in the event was empty.

Tesla was missing!

The side hall, leading off the stage, and the anterooms were searched, but there was no sign of him. Members of the committee slipped out to retrace their steps through the lobby and back to the Club dining room. A man as tall as Tesla could not be hidden in any group, yet there was not a sign of him in either building.

The delay in opening the meeting in the auditorium was embarrassing—but the ceremonies could not be started without Tesla, and where was he?

It seemed hardly possible that an imposing figure like Tesla, his height exaggerated by the streamlined contours of his swallow-tailed formal evening dress, and in the almost worshipful custody of a score of outstanding intellects, could vanish without any of them observing his going.

Behrend rushed back from the Club to the auditorium, hopeful that Tesla had preceded him; but he found that such was not the case. All the washrooms in both buildings had been searched; he was concealed in none of them. No one could offer a theory to account for his disappearance.

None but Behrend knew of Tesla's aversion to accepting the

Edison Medal, yet even he had not the slightest knowledge of what had become of the famous inventor. He recalled noting the shadowy walks of Bryant Park opposite the Club as he and Tesla stepped from the taxicab earlier in the evening, and he wondered if Tesla had retreated there for some quiet meditation before the ceremony. He hurried out of the Club.

As Behrend stepped into Bryant Park, the last faint glimmerings of dusk were visible in the high sky; but in the park the shades of night were gathering and here and there could be heard the faint twitterings of birds. The twittering of the birds brought, like a flash, to Behrend's mind the scene he had observed in Tesla's apartment at the Hotel St. Regis. In the room which Tesla had arranged as a reading room and office was a roll-top desk, and on top of this were four neat circular baskets, in two of which pigeons were nestled. Before they left the apartment Tesla went to the window, which was kept open at all times, whistled softly, and two more pigeons quickly flew into the room. Just before leaving for the dinner Tesla fed the pigeons, and having done so slipped a paper bag filled with something into his pocket. The possible significance of this latter act did not occur to Behrend until he heard the twittering of the birds in the park.

With all possible speed Behrend rushed out of the park, down 40th Street toward Fifth Avenue, and up the steps to the plaza of the Library. Here he beheld a sight that amazed him almost beyond belief in what his eyes told him. Here was the missing man. He had recalled that Tesla regularly visited the Library, St. Patrick's Cathedral, or other places to feed the pigeons.

In the center of a large thin circle of observers stood the imposing figure of Tesla, wearing a crown of two pigeons on his head, his shoulders and arms festooned with a dozen more, their white or pale-blue bodies making strong contrast with his black suit and black hair, even in the dusk. On either of his outstretched hands was another bird, while seemingly hundreds more made a living carpet on the ground in front of him, hopping about and pecking at the bird seed he had been scattering.

It was Behrend's impulse to rush in, shoo the birds away and, seizing the missing man, rush him back to the auditorium. Something caused him to halt. Such an abrupt action seemed almost sacrilegious. As he hesitated momentarily, Tesla caught sight of him and slowly shifted the position of one hand to raise a warning finger. As he did so, however, he moved slowly toward Behrend; and as he came close, some of the birds flew from Tesla's shoulders to Behrend's. Apparently sensing a disturbing situation, though, all the birds flew to the ground.

Appealing to Tesla not to let him down, nor to embarrass those who were waiting at the meeting, Behrend prevailed upon the inventor to return to the auditorium. Little did Behrend know how much more the pigeons meant to Tesla than did the Edison Medal; and little could anyone have suspected the fantastic secret in Tesla's life, of which the outer manifestation was his faithful feeding of his feathered friends. To Behrend it was just another, and in this case very embarrassing, manifestation of the nonconformity of genius. Of this, more later.

Returning to the auditorium, Behrend explained in a quick aside to the president that Tesla had been temporarily ill, but that his condition was now quite satisfactory. The opening of the meeting had been delayed about twenty minutes.

In his presentation speech, Behrend pointed out that by an extraordinary coincidence, it was exactly 29 years ago, to the very day and hour, that Nikola Tesla presented his original description of his polyphase alternating-current system. He added:

Not since the appearance of Faraday's "Experimental Researches in Electricity" has a great experimental truth been voiced so simply and so clearly as this description of Mr. Tesla's great discovery of the generation and utilization of polyphase alternating currents. He left nothing to be done by those who followed him. His paper contained the skeleton even of the mathematical theory.

Three years later, in 1891, there was given the first great demonstration, by Swiss engineers, of the transmission of power at 30,000 volts from Lauffen to Frankfort by means of Mr. Tesla's system. A few years later this was followed by the development of the Cataract Construction Company, under the presidency of our member, Mr. Edward D. Adams, and with the aid of the engineers of the Westinghouse Company. It is interesting to recall here tonight that in Lord Kelvin's report to Mr. Adams, Lord Kelvin recommended the use of direct current for the development of power at Niagara Falls and for its transmission to Buffalo.

The due appreciation or even enumeration of the results of Mr. Tesla's invention is neither practicable nor desirable at this moment. There is a time for all things. Suffice it to say that, were we to seize and eliminate from our industrial world the results of Mr. Tesla's work, the wheels of industry would cease to turn, our electric cars and trains would stop, our towns would be dark, our mills would be dead and idle. Yes, so far reaching is this work, that it has become the warp and woof of industry. . . . His name marks an epoch in the advance of electrical science. From that work has sprung a revolution in the electrical art.

We asked Mr. Tesla to accept this medal. We did not do this for the mere sake of conferring a distinction, or of perpetuating a name; for so long as men occupy themselves with our industry, his work will be incorporated in the common thought of our art, and the name of Tesla runs no more risk of oblivion than does that of Faraday, or that of Edison.

Nor indeed does this Institute give this medal as evidence that Mr. Tesla's work has its official sanction. His work stands in no need of such sanction.

No, Mr. Tesla, we beg you to cherish this medal as a symbol of our gratitude for a new creative thought, the powerful impetus, akin to revolution, which you have given to our art and to our science. You have lived to see the work of your genius established. What shall a man desire more than this? There rings out to us a paraphrase of Pope's lines on Newton:

"Nature and Nature's laws lay hid in night:
"God said, Let Tesla be, and all was light."

No record remains of Tesla's acceptance speech. He did not prepare a formal address. He had intended to make but a brief response, but instead he became involved in anecdotal narration and a preview of the future of electrical science which, in the absence of the limiting influence of a written copy, became quite lengthy.

It is doubtful if anyone in the audience, or on the stage, grasped the full significance of Behrend's words when he said, "We asked Mr. Tesla to accept this medal." And fewer still were the members of the Institute who had any conception of the extent or importance of Tesla's contribution to their science. His major inventions had been announced thirty years before. The majority of the engineers present belonged to the younger generation; and they had been taught from textbooks that almost completely omitted mention of Tesla's work.

## FIFTEEN

THE announcement by Tesla in his latter years that attracted the greatest amount of attention concerned his discovery of what has briefly, but not too accurately, been termed a death ray. Earlier reports had come from Europe of the invention of death rays, beams of radiation that would cause airships on which they impinged to burst into flame, the steel bodies of tanks to melt and the machinery of ships to stop operating, but all gave indications of being part of the game of diplomatic buncombe.

The prelude to Tesla's death-ray announcement came several years in advance, in the form of a declaration that he had made discoveries concerning a new form of power generation which, when applied, would make the largest existing turbine-dynamo units in the powerhouses look like pygmies. He made this announcement in interviews with the press in 1933, and declared that he was also working on a new kind of generator for the production of radiation of all kinds and in the greatest intensities. He made similar announcements the following year.

237

Both of these announcements were entitled to receive the most serious consideration, even though they were not accompanied by experimental evidence, and revealed no technical details.

When Tesla was talking as a scientist he was opposed to wars on moral, economic and all practical and theoretical grounds. But, like most scientists, when he stopped thinking as a scientist and let his emotions rule his thoughts, he found exceptions in which he felt some wars and situations were justifiable. As a scientist he was unwilling to have the discoveries of scientists applied to the purposes of war makers, but when the emotional phase of his nature took the ruling position he was willing to apply his genius to devising measures that would prevent wars by supplying protective devices.

This attitude is exemplified in the following statement, which he had prepared in the twenties but did not publish:

At present many of the ablest minds are trying to devise expedients for preventing a repetition of the awful conflict which is only theoretically ended and the duration and main issues of which I correctly predicted in an article printed in the Sun of December 20, 1914. The League is not a remedy but, on the contrary, in the opinion of a number of competent men, may bring about results just the opposite. It is particularly regrettable that a punitive policy was adopted in framing the terms of peace because a few years hence it will be possible for nations to fight without armies, ships or guns, by weapons far more terrible, to the destructive action and range of which there is virtually no limit. Any city at any distance whatsoever from the enemy can be destroyed by him and no power on earth can stop him from doing so. If we want to avert an impending calamity and a state of things which may transform this globe into an inferno, we should push the development of flying machines and wireless transmission of energy without an instant's delay and with all the power and resources of the nation.

Tesla saw preventive possibilities in his new invention which embodied "death-ray" characteristics, and which was made several years after the foregoing statement was written. He saw it providing a curtain of protection which any country, no matter how small, could use as a defense against invasion. While he might offer it as a defensive weapon, however, there would be nothing to stop military men from using it as a weapon of offense.

Tesla never gave the slightest hint concerning the principles under which his device operated.

THERE are indications, at any rate, that Tesla was working on a high-potential direct-current system for generating and transmitting electricity to long distances. Direct current at very high voltages can be transmitted much more efficiently than alternating current. There has been no practical way of generating direct current at high voltages. It was because of this that Tesla's polyphase alternating-current system was adopted for our present nationwide superpower system, since it made the use of high voltages practicable. But, despite its efficiencies, it entailed certain losses which could be eliminated if direct current of sufficiently high voltage could be obtained. Such a system would supersede his alternating-current system but not displace it.

Direct current, perhaps at several million volts potential, would be used to transmit current for long distances, perhaps clear across the continent, providing a kind of express transmission system, to which the existing alternating-current system would be tied for local distribution. In addition to the direct-current transmission system, he appears to have worked out a high-voltage direct-current generator and a new type of direct-current motor which would operate without a commutator.

The inventions were starting to dam up in Tesla's mind like water in a reservoir to which there was no outlet.

Just as he developed his alternating-current system into the high-frequency, high-potential field of power distribution by wireless, which he demonstrated at Colorado Springs, so he appears to have carried his direct-current system forward and linked it with his alternating-current wireless distribution system, so that he could use both in a super-interlocking system. As this remained unapplied, he further evolved it and produced a plan for operating with it what appears to be a beam system of wireless transmission of energy which might involve the use of a stream of particles such as are used in the atom-smashing cyclotron.

As time passed from the latter twenties, through the latter thirties, the hints which Tesla would drop about his work became more complicated, and so ambiguous that they aroused skepticism rather than respect. He would not reveal the nature of his discoveries until he had secured patents, and he would not apply for patents until he had made actual working models, and he could not make the working models because he had no money. Samuel Insull, the public utilities magnate, had for many years made frequent and generous contributions to Tesla. They were usually applied to outstanding debts and were not large enough to enable him to engage in laboratory research work.

Tesla, however, never exhibited the slightest outward sign of bitterness over the situation. Instead, he always appeared in the rôle of confirmed optimist, always maintaining a spirit of hopefulness that he would achieve by his own efforts the money he needed to carry out his elaborated plans. This is indicated in a letter he wrote to B. A. Behrend, who had induced him to accept the Edison Medal, and who was probably in his confidence to a greater extent than anyone else:

"I am hard at work on those discoveries of mine, I told you about, from which I hope to derive a sum in eight figures (not counting the cents, of course) enabling me to erect that wireless power plant at my own expense. And what I shall accomplish by that other invention I came specially to see you about, I do not dare to tell you. This is stated in all seriousness."

The invention about which he dared not speak was probably his direct-current generating and transmitting system.

In an interview given in 1933, he said his power generator was of the simplest kind—just a big mass of steel, copper and aluminum, comprising a stationary and a rotating part, peculiarly assembled. He was planning, he said, to generate electricity and transmit it to a distance by his alternating system; but the direct-current system could also be employed if the heretofore insuperable difficulties of insulating the transmission line could be overcome.

A year later he had developed the beam-transmission plan; and he made an ambiguous statement concerning it which was reported in the press as news of a "death ray" since the description seemed to fit into the same mold as those wild and improbable statements that had come out of Europe some years before. A writer in the New York *World-Telegram* described Tesla's plan as "nebulous." This drew a reply from Tesla (July 24, 1934) in which the following paragraphs appeared:

Still another item which has interested me is a report from Washington in the World Telegram of July 13, 1934, to the effect that scientists doubt the death ray effects. I am quite in agreement with these doubters and probably more pessimistic in this respect than anybody else, for I speak from long experience.

Rays of the requisite energy cannot be produced, and, then again, their intensity diminishes with the square of the distance. Not so the agent I employ, which will enable us to transmit to a distant point more energy than is possible by any other kind of ray.

We are all fallible, but as I examine the subject in the light of my present theoretical and experimental knowledge I am filled with deep convictions that I am giving to the world something far beyond the wildest dreams of inventors of all time.

This is the first written statement by Tesla in which he mentions his "ray"; but I had, as already noted, obtained some confidential statements from him, during the preceding year or so, concerning results he hoped to achieve through his new discovery, the nature of which he kept as a well-protected secret. Three years later, in 1937, Tesla permitted me to write a news story for the New York *Herald Tribune* on his new power-and-ray discovery. In it I stressed the usefulness of the discovery for delivering power to ships for travel across the ocean, thus eliminating the need for carrying fuel supplies, rather than its use as a weapon for defense or offense.

On this occasion I tried to get him to reveal some technical details, but he successfully parried every question and gave no information beyond the statement that the transmitting plant on shore was one which he would be able to erect at a cost of about $2,000,000, and the energy would be transmitted by a ray or beam of infinitesimally small cross section, one hundred thousandth of a centimeter in diameter. To other newspapers which copied my story he gave the figure as one millionth of a square centimeter.

Later, I wrote a somewhat critical review of his plan and sought to draw him out by reviewing the properties of electromagnetic radiation in all parts of the spectrum. Finding none that possessed any known characteristics needed to make his ray practical, I also reviewed the properties of all known particles of matter, and stated that none of these would serve his purpose with the possible exception of the unelectrified particle, the neutron. He made no revealing response to the article.

At his birthday dinner in 1938, at the Hotel New Yorker, Tesla described briefly his combination wireless-power transmission and death ray, adding little to what has already been stated; and in a later part of his speech he declared that he had developed a method for interplanetary communication, in which he would be able to transmit not only communication signals of small strength but energies involving thousands of horsepower.

On this occasion I asked him if he would be specific concerning the effects produced, and whether they would be visible from the earth; for example—could he produce an effect on the moon sufficiently large to be seen by an astronomer watching the moon through a high-power telescope? To this he replied that he would be able to produce in the dark region of the thin crescent new moon an incandescent spot that would glow like a bright star so that it could be seen without the aid of a telescope.

It would appear probable that Tesla proposed to use for this purpose the beam he described in connection with his wireless-power "death ray." The limitation of the destructive effects of the beam, which he visualized as two hundred miles, was due to the fact that the beam had a straight-line trajectory. Tesla stated that the curvature of the earth set a limit on the distance of operation, so the two-hundred-mile span of operation gave an indication of the greatest practical height of a tower from which the beam could be directed. He expected to use potentials of about 50,000,000 volts in his system, but whether of direct or alternating current is unknown.

The only written statement by Tesla on this subject is in his manuscript of the talk which was delivered, in absentia, some months later before the Institute of Immigrant Welfare in response to its honorary citation. In this was included the following paragraph:

"To go to another subject: I have devoted much of my time during the year past to perfecting of a new small and compact apparatus by which energy in considerable amounts can now be flashed through interstellar space to any distance without the slightest dispersion. I had in mind to confer with my friend, George E. Hale, the great astronomer and solar expert, regarding the possible use of this invention in connection with his own researches. In the meantime, however, I am expecting to put before the Institute of France an accurate description of the device with data and calculations and claim the Pierre Gutzman Prize of 100,000 francs for means of communication with other worlds, feeling perfectly sure that it will be awarded to me. The money, of course, is a trifling consideration, but for the great historical honor of being the first to achieve this miracle I would be almost willing to give my life."

243

# FOURTH PART

## SELF-MADE SUPERMAN

# SIXTEEN

I T WAS during a period when he was most busily occupied with his experiments with high-frequency and high-potential currents, from 1892 to 1894, that Tesla had found time to give serious thought to yet another type of problem, matter and energy; and from it he derived what he described as a new physical principle. This he developed to the point at which he was able to propound a new dynamic theory of gravity.

While this principle guided much of his thinking, he did not make any announcements concerning it until close to the end of his life. Such disclosures as have been made, however, leave this much obvious: Tesla considered his theory wholly inconsistent with the theory of relativity, and with the modern theory concerning the structure of the atom and the mutual interconversion of matter and energy. Tesla continuously attacked the validity of Einstein's work; and until two or three years before his death, he ridiculed the belief that energy could be obtained from matter.

These antagonisms were most unfortunate, as they placed Tesla in conflict with modern experimental physics. This was totally unnecessary, for Tesla could undoubtedly have adhered to his principle and interpreted it so that it was not inconsistent with the modern theories. The antagonism was probably attributable to psychological factors rather than scientific inconsistencies.

The only statement Tesla has made concerning his principle and his theory is that contained in the lecture he prepared for delivery before the Institute of Immigrant Welfare (May 12, 1938). In this he stated:

During the succeeding two years [1893 and 1894] of intense concentration I was fortunate enough to make two far reaching discoveries. The first was a dynamic theory of gravity, which I have worked out in all details and hope to give to the world very soon. It explains the causes of this force and the motions of heavenly bodies under its influence so satisfactorily that it will put an end to idle speculation and false conceptions, as that of curved space. . . .

Only the existence of a field of force can account for the motions of the bodies as observed, and its assumption dispenses with space curvature. All literature on this subject is futile and destined to oblivion. So are all attempts to explain the workings of the universe without recognizing the existence of the ether and the indispensable function it plays in the phenomena.

My second discovery was of a physical truth of the greatest importance. As I have searched the entire scientific records in more than a half dozen languages for a long time without finding the least anticipation, I consider myself the original discoverer of this truth, which can be expressed by the statement: There is no energy in matter other than that received from the environment.

On my 79th birthday I made a brief reference to it, but its meaning and significance have become clearer to me since then. It applies rigorously to molecules and atoms as well as to the largest heavenly bodies, and to all matter in the universe in any phase of its existence from its very formation to its ultimate disintegration.

Tesla's mind was inflexible in the matter of his attitude toward relativity and the modern theories. Had he published his principle and theory of gravity at the beginning of the century it would, without doubt, have then received very serious consideration and perhaps general acceptance, although it is difficult to make an intelligent surmise without knowledge of his postulates. If published, it might have had some influence on Einstein's thinking. The field of force which Tesla mentions as being necessary to explain the movements of the planets might have been his contribution to eliminating the need for the ether which was accomplished by Einstein's theory. The two theories might have been merged, in which case there probably would have resulted a harmonious development of the thinking of the two geniuses.

In this latter case Tesla could very well have shaped his thinking to see a consistency between his theory that there is no energy in matter other than that received from its environment, and the modern viewpoint that all matter consists of energy into which it is convertible; for when matter is converted to energy, the energy returns to the environment from whence it came when the particles were formed.

There appears to be a frustration involved in Tesla's attitude which could have been resolved by early publication of his theory. If this had taken place, Tesla's powerful intellect and his strange ability to solve problems would have been brought to bear on the problems of atomic physics and he, in turn, would have received tremendous benefits from the application of the newer knowledge in the fields in which he was supreme.

Tesla's ability to generate tremendously high voltages would have been of great assistance in the task of "smashing the atom." Other scientists, even today, are struggling to produce currents with a potential of 5,000,000 volts, whereas Tesla, forty years ago, had generated potentials of 135,000,000 volts.

The inconsistency between Tesla's principle and the picture of the atom consisting of a small complex nucleus surrounded by planetary electrons—which inconsistency was more existent in Tesla's mind than in Nature—caused him to develop an antagonism to all scientific developments which called for a picture that differed from the billiard-ball type of atom in vogue in the eighteen-eighties. To him, a smashed atom was like a smashed billiard ball.

The electron, however, had a real existence to Tesla. He accepted it as a kind of sub-atom, a fourth state of matter, as described by Sir William Crookes, who discovered it. Tesla visualized it as associated with but not a part of the atom. The electric charge it carried was entirely distinct from the electron. Electricity, to him, was a fluid much more highly attenuated than any known form of matter, and with highly specific properties of its own for which it was not dependent upon matter. The charge on the electron was due to a surface layer of electricity covering it, and it could receive many layers, giving it multiple charges, all of which could be dissipated. These statements were similar to those which he had published a half-century before.

According to the modern theory, on the other hand, the electrical nature of the electron, described as its charge, is a characteristic inherent in the nature of the energy crystallized about a point which gives the electron its existence, and the electron is one of the particles, or units of energy, of which the atom is composed.

In discussing articles by scientists in the field of atomic physics, Tesla would register his protests that their theories were untenable and the claims unfounded; and he was particularly emphatic when experiments in which energy emissions from atoms were recorded.

"Atomic power is an illusion," he frequently declared. He furnished several written statements in which he said that with his currents of several million volts he had, countless times, smashed uncounted billions of atoms—and he knew that no emission of energy accompanied the process.

On one occasion Tesla took me to task rather severely for my failure to publish his statements. I replied: "I withheld them in order to protect your reputation. You are making too great a virtue of consistency. It is not necessary that you adhere to the theories you held as a youth, and I am convinced that deep down in your heart you hold newer theories that are in harmony with scientific developments in other fields, but because you have disagreed with, and attacked, some modern theories you feel you must be consistent and attack them all. I am convinced that in the development of your death-ray device your thinking was along the lines of the modern theory of the structure of the atom and the nature of matter and energy."

Tesla thereupon let me know in no uncertain terms that he held very definite ideas concerning efforts on the part of others to do his thinking for him. This conversation took place about 1935; and I did not hear from him for many months. I observed, however, that in his later conversations he was much less dogmatic concerning modern theories, and a few years later he stated that he was planning an apparatus which would make possible a definite testing of the modern theory of atomic structure, with the expectation that his new power system and energy beam would release atomic energy more effectively than any device then in use by physicists.

Having endorsed, finally, the belief that man will be able to smash, transmute, create or destroy atoms, and control vast amounts of energy, he waxed poetic on the subject. He extended man's control over atoms and energy to a cosmic scale, and saw him shaping the universe according to our desires. In an unpublished article, entitled "Man's Greatest Achievement," he wrote:

There manifests itself in the fully developed being—Man—a desire mysterious, inscrutable and irresistible: to imitate nature, to create, to work himself the wonders he perceives. Inspired to this task he searches, discovers and invents, designs and constructs, and covers with monuments of beauty, grandeur and awe, the star of his birth. He descends into the bowels of the globe to bring forth its hidden treasures and to unlock its immense imprisoned energies for his use. He invades the dark depths of the ocean and the azure regions of the sky. He peers into the innermost nooks and recesses of molecular structure and lays bare to his gaze worlds infinitely remote. He subdues and puts to his service the fierce, devastating spark of Prometheus, the titanic forces of the waterfall, the wind and the tide. He tames the thundering bolt of Jove and annihilates time and space. He makes the great Sun itself his obedient toiling slave. Such is his power and might that the heavens reverberate and the whole earth trembles by the mere sound of his voice.

What has the future in store for this strange being, born of a breath, of perishable tissue, yet immortal, with his powers fearful and divine? What magic will be wrought by him in the end? What is to be his greatest deed, his crowning achievement?

Long ago he recognized that all perceptible matter comes from a primary substance, or a tenuity beyond conception, filling all space, the Akasa or luminiferous ether, which is acted upon by the life-giving Prana or creative force, calling into existence, in never ending cycles, all things and phenomena. The primary substance, thrown into infinitesimal whirls of prodigious velocity, becomes gross matter; the force subsiding, the motion ceases and matter disappears, reverting to the primary substance.

Can Man control this grandest, most awe-inspiring of all processes in nature? Can he harness her inexhaustible energies to perform all their functions at his bidding, more still cause them to operate simply by the force of his will?

If he could do this, he would have powers almost unlimited and supernatural. At his command, with but a slight effort on his part, old worlds would disappear and new ones of his planning would spring into being. He could fix, solidify and preserve the ethereal shapes of his imagining, the fleeting visions of his dreams. He could express all the creations of his mind on any scale, in forms concrete and imperishable. He could alter the size of this planet, control its seasons, guide it along any path he might choose through the depths of the Universe. He could cause planets to collide and produce his suns and stars, his heat and light. He could originate and develop life in all its infinite forms.

To create and to annihilate material substance, cause it to aggregate in forms according to his desire, would be the supreme manifestation of the power of Man's mind, his most complete triumph over the physical world, his crowning achievement, which would place him beside his Creator, make him fulfill his ultimate destiny.

Tesla, in his eighties, was still manifesting the superman complex, and on even more elaborate a scale than when in his twenties. In his earlier dreams his visions were terrestrial, but in later life they were extended to embrace the entire universe.

Even on the cosmic scale, however, Tesla spoke in terms of matter and energy. These two entities, according to his reasoning, were sufficient to explain all observed phenomena, a situation which militated against the discovery of any new agencies.

The civilizations of the ancient world knew nothing of electricity and magnetism; the controlled manifestations of these two phases of a single force-entity have provided us with a new civilization and a new cultural outlook on life, as well as broadened horizons within the life sphere. There is no reason why we should not look forward to the discovery of new forces which are as different from electricity as electricity is from the winds of the air and the waves of the ocean. If inadequate explanations of vital phenomena are accepted as satisfactory, embracing extravagant extensions of known forces, the way is closed to the discovery of unknown forces and the opening of any new realms of knowledge. This was the limitation which the science of the last quarter of the past century placed upon itself; and Tesla's philosophy was a product of that period. It was difficult for him to reshape it in his later years.

THE memory departments of most individuals' brains are like office filing systems, an excellent dumping ground for everything that comes along—but try to find a filed item later. Tesla's powers of memorizing were prodigious. A quick reading of a page gave him a permanent record of it; he could always recall before his eyes a photographic record of it to be read, and could study at his convenience. Study, for Tesla, was a far different process than for the average person. He had no need for a reference library; he could consult in his mind any page of any textbook he had read, any formula, equation, or item in a table of logarithms, and it would flash before his eyes. He could recite scores of books, complete from memory. The saving in time which this made possible in research work was tremendous.

This strange faculty of vision was supernormal but entirely natural and was due, probably, to a structural characteristic in his brain which provided a direct channel between the memory and the visual areas of his cerebral hemispheres. It provided him with a very useful new sense.

The human brain is made up of two sections, the right and left sides, each of which, in some of its phases, is a complete brain; and both halves function together as a single unit. There are many layers in the brain parallel with its surface, each connected to the others by complex nerve fibers, like threads sewing together the layers of an onion. The outer layer seems to be directly associated with our consciousness. The surface is divided into specialized areas. There is a band across the mid-section of each hemisphere, from ear to ear over the top of the brain, devoted to the senses, and here are separate areas for the sensory faculties—sight, hearing, taste, smell—while near by are regions for the motor or muscular activities of the various parts of the body. The back lobe of the brain appears to be the home of the memory and the front lobe of some higher faculties of integration, the nature of which we do not as yet understand.

In normal processes of seeing, the eye forms a picture of an object on the retina, a screen on the back of the eyeball. The retina is supplied with thousands of nerve endings all packed together like stalks of asparagus in a bunch. The tip ends are provided with photosensitive processes, and when light strikes any one of them it transmits over the optic nerve a signal to the brain which is recorded as a visual response in the sight area of each half of the brain. By co-operation of all the nerve endings, the complete picture seen is transmitted. The actual seeing, then, is done in the brain and not in the eye. When an object is seen by the brain, a record of that visual experience is transmitted from the sight area of the brain to the memory center in the back part of the brain; and similar records are sent by all other sensory centers. Ordinarily this is a one-way process, the stimuli going in the direction of the memory and nothing coming back to the sensory area. If this were not so, our sense areas of the brain would be continuously re-enacting old experiences and mixing them with the new, incoming experiences, causing annoying confusion.

The memory area contains a complete record of all sensory experiences we have had. In our thinking processes we use some little-understood mechanism for connecting together items stored in the memory area to produce useful combinations or relationships, or, in other words, new ideas. The memory appears to function on a subconscious level but we seem to be able to activate fibers that reach down to the desired strata at the right point to connect the memory level with the consciousness level. In this way we can recall experiences, but this experience of memory is far different from the original experience of sight out of which the original memory record was made.

If, however, in this process of recollection, the nerve fiber linking the sight area of the brain and the memory area were to be activated, then we would see again by the sharp processes of vision the object which caused the memory record we are trying to recollect.

The act of creative thinking seems to consist of assembling two or more memory records of sensory experiences into a combination which possesses entirely new characteristics that were not apparent in the component parts. If the nerve linkage just referred to were to operate in a two-way process with the visual area, then we would be able to see the new creation as if it were a really existing object seen by the eye, although the whole operation was limited to the brain.

This process is hypothetically the one which took place in Tesla's brain and gave him tremendously greater powers of creative work than are possible to the ordinary individual. Was this conceivably a new invention made by Mother Nature and tried out by her on Tesla?

Tesla himself never understood the neurological, or physiological, processes underlying this strange faculty. To him it was an absolutely real experience to see in front of him as solid objects the subjects of his creative thoughts. He believed that the image of the thing he saw was sent back from the brain along the optic nerve to the eye, and that it existed as a picture on the retina where, by some suitable means, it could be seen by others— or that by means of adequate amplifying devices, such as are used in television, it could be projected on a screen. He even proposed such devices.*

Tesla described his experience with this strange faculty in an

---

* The apparent flaw in his reasoning followed on his mistake in thinking that he was doing this supernormal seeing with his eye, whereas the process was confined to his brain; and the reflex action from the memory centers stopped at the visual centers instead of, as he believed, being continued forward through the optical nerve to the retina.

interview with M. K. Wisehart, published under the title "Making Your Imagination Work for You" in the *American Magazine*, April, 1921. He stated:

During my boyhood I had suffered from a peculiar affliction due to the appearance of images, which were often accompanied by strong flashes of light. When a word was spoken, the image of the object designated would present itself so vividly to my vision that I could not tell whether what I saw was real or not. . . . Even though I reached out and passed my hand through it, the image would remain fixed in space.

In trying to free myself from these tormenting appearances, I tried to concentrate my thoughts on some peaceful, quieting scene I had witnessed. This would give me momentary relief; but when I had done it two or three times the remedy would begin to lose its force. Then I began to take mental excursions beyond the small world of my actual knowlege. Day and night, in imagination, I went on journeys—saw new places, cities, countries, and all the time I tried hard to make these imaginary things very sharp and clear in my mind. I imagined myself living in countries I had never seen, and I made imaginary friends, who were very dear to me and really seemed alive.

This I did constantly until I was seventeen, when my thoughts turned seriously to invention. Then, to my delight, I found I could *visualize* with the greatest facility. I needed no models, drawings, or experiments. I could picture them all in my mind. . . .

By that faculty of *visualizing*, which I learned in my boyish efforts to rid myself of annoying images, I have evolved what is, I believe, a new method of materializing inventive ideas and conceptions. It is a method which may be of great usefulness to any imaginative man, whether he is an inventor, businessman or artist.

Some people, the moment they have a device to construct or any piece of work to perform, rush at it without adequate preparation, and immediately become engrossed in details, instead of the central idea. They may get results, but they sacrifice quality.

Here, in brief, is my own method: After experiencing a desire to invent a particular thing, I may go on for months or years with the idea in the back of my head. Whenever I feel like it, I roam around in my imagination and think about the problem without any deliberate concentration. This is a period of incubation.

Then follows a period of direct effort. I choose carefully the possible solutions of the problem. I am considering, and gradually center my mind on a narrowed field of investigation. Now, when I am deliberately thinking of the problem in its specific features, I may begin to feel that I am going to get the solution. And the wonderful thing is, that if I do feel this way, *then I know I have really solved the problem and shall get what I am after.*

The feeling is as convincing to me as though I already had solved it. I have come to the conclusion that at this stage the actual solution is in my mind *subconsciously,* though it may be a long time before I am aware of it *consciously.*

Before I put a sketch on paper, the whole idea is worked out mentally. In my mind I change the construction, make improvements, and even operate the device. Without ever having drawn a sketch I can give the measurements of all parts to workmen, and when completed all these parts will fit, just as certainly as though I had made the actual drawings. It is immaterial to me whether I run my machine in my mind or test it in my shop.

The inventions I have conceived in this way have always worked. In thirty years there has not been a single exception. My first electric motor, the vacuum tube wireless light, my turbine engine and many other devices have all been developed in exactly this way.

That Tesla believed his mental visualizations brought images from his brain to the back of his eye is indicated by some statements he made in his famous lecture before the National Electric Light Association convention at St. Louis, in March, 1893, when announcing his discovery of radio. These statements about vision had no relationship to the subject of the lecture, and the fact that he interjected them indicated that his experiences with this strange power had a powerful influence on his inventive thinking. He said:

It can be taken as a fact, which the theory of the action of the eye implies, that for each external impression, that is for each image produced on the retina, the ends of the visual nerves, concerned in the conveyance of the impressions to the mind, must be under a peculiar stress or in a vibratory state. It now does not seem improbable that, when by the power of thought an image is evoked, a distinct reflex action, no matter how weak, is exerted upon certain ends of the visual nerves, and therefore upon the retina. Will it ever be within human power to analyze the condition of the retina, when

257

disturbed by thought or reflex action, by the help of some optical or other means of such sensitiveness that a clear idea of its state might be obtained? If this were possible, then the problem of reading one's thoughts with precision, like the characters of an open book, might be much easier to solve than many problems belonging to the domain of positive physical science, in the solution of which many, if not the majority, of scientific men implicitly believe.

Helmholtz has shown that the fundi of the eye are themselves luminous, and he was able to see in total darkness the movements of his arm by the light of his own eyes. This is one of the most remarkable experiments recorded in the history of science, and probably only a few men could satisfactorily repeat it, for it is very likely that the luminosity of the eyes is associated with uncommon activity of the brain and great imaginative power. It is fluorescence of brain action, as it were.

Another fact having a bearing on this subject, which has probably been noted by many, since it is stated in popular expressions, but which I cannot recollect to have found chronicled as a positive result of observation is that, at times, when a sudden idea or image presents itself to the intellect, there is a painful sensation of luminosity produced in the eye observed even in broad daylight.

Forty years later Tesla was still interested in the possibility of capturing a photographic record of thoughts. He stated in interviews that if his theory were correct—that thoughts are recorded on the retina—it should be possible to photograph what is revealed on this screen in the eye, and project enlarged images of it.

There is nothing illogical about Tesla's reasoning concerning his strange faculty of visualizing and the possibility of finding a corresponding image on the retina. There is a bare possibility that in an extreme case, as was his, a reflex arc may have extended from the brain to the retina; but the probability that it did not is stronger. If he had possessed the ability to take others into his confidence in his experiments, he would have been able to stage some simple tests in the laboratory of an ophthalmologist which would have given him some definite experimental evidence to support or dispose of his theories, as far as photographic thought images were concerned.

ABOUT 1920 Tesla had prepared, although he never published, an announcement of what he declared was "An Astounding Discovery." It involved factors which he called "cosmic"; but it likewise presented situations which the practicers of voodoo in Hayti, and other intellectually unveneered portions of the human race, would receive with perfect understanding. Since Tesla, one of the most highly civilized individuals, could evolve this conception, it is probable that other supercultured individuals or groups could find it in harmony with their ideas and experiences.

It involves, however, a situation in which the soulless "matter and energy" automaton (to which status we have seen Tesla relegate human beings) is able to judge ethical values, and, like a pontiff presiding over a court of morals, inflict punishment for transgressions.

Here is Tesla's description of his "astounding discovery":

While I have failed to obtain any evidence in support of the contentions of psychologists and spiritualists, I have proved to my complete satisfaction the automatism of life, not only through continuous observation of individual actions, but even more conclusively, through certain generalizations. These amount to a discovery which I consider of the greatest moment to human society and on which I shall briefly dwell.

I got the first inkling of this astounding truth when I was still a very young man, but for many years I interpreted what I noted simply as coincidences. Namely, whenever either myself or a person to whom I was attached, or a cause to which I was devoted, was hurt by others in a particular way, which might be best popularly characterized as the most unfair imaginable, I experienced a singular and undefinable pain which, for want of a better term, I have qualified as "cosmic," and shortly thereafter, and invariably, those who have inflicted it came to grief. After many such cases I confided this to a number of friends, who had the opportunity to convince themselves of the truth of the theory which I have gradually formulated and which may be stated in the following words.

Our bodies are of similar construction and exposed to the same external influences. This results in likeness of response and concordance of the general activities on which all our social and other rules and laws are based. We are automata entirely controlled by the forces of the medium, being tossed about like corks on the surface of the water, but mistaking the resultant of the impulses from the outside for free will.

The movements and other actions we perform are always life-preservative and though seemingly quite independent from one another, we are connected by invisible links. So long as the organism is in perfect order it responds accurately to the agents that prompt it, but the moment there is some derangement in any individual, his self-preservative power is impaired.

Everybody understands, of course, that if one becomes deaf, has his eyesight weakened, or his limbs injured, the chances for his continued existence are lessened. But this is also true, and perhaps more so, of certain defects in the brain which deprive the automaton, more or less, of that vital quality and cause it to rush into destruction.

A very sensitive and observant being, with his highly developed mechanism all intact, and acting with precision in obedience to the changing conditions of the environment, is endowed with a transcending mechanical sense, enabling him to evade perils too subtle to be directly perceived. When he comes in contact with others whose controlling organs are radically faulty, the sense asserts itself and he feels the "cosmic" pain.

The truth of this has been borne out in hundreds of instances and I am inviting other students of nature to devote attention to this subject, believing that, through combined and systematic effort, results of incalculable value to the world will be attained.

Tesla's uncommunicative nature concerning his own intimate experiences has undoubtedly deprived the world of many interesting stories. He was unquestionably an abnormal individual, and of a type that does have what are known as "psychic experiences." He was emphatic in his denial that he ever had experiences of that sort; yet he has related incidents that clearly belong in the psychic category. He seemed to be fearful that an admission of psychic experiences would cause him to be misunderstood as supporting spiritualism, or theories that something operates in life other than matter and energy.

Whenever he was asked for his philosophy of life, he would

elaborate a theory that the human body is a meat machine which responds to external forces.

One evening in New York, as Tesla and the author sat in the lobby of the Hotel Governor Clinton, the inventor discussed his meat-machine theory. It was a materialistic philosophy typical of the Victorian era. We are, he held, composed of only those things which are identified in the test tube and weighed in the balance. We have only those properties which we receive from the atoms of which our bodies are constructed. Our experiences, which we call life, are a complex mixture of the responses of our component atoms to the external forces of our environment.

Such a philosophy has the virtue of simplicity and brevity of presentation; and it lends itself readily to being propounded with a positiveness that reacts on the propounder, and transforms his attitude into one of dogmatism in which emphatically expressed opinion is often confused with and substituted for factual evidence.

"I don't believe a word of your theory," I replied to Tesla's exposition, "and, thank God, I am convinced you don't believe a word of it either. The strongest proof I have that your theory is totally inadequate is that Tesla exists. Under your theory we could not have a Tesla. Tesla possesses a creative mind and, in his accomplishments, stands high above all other men. If your theory were correct, we would either all be geniuses like Tesla or we would all be mental mediocrities living in these meat machines you describe, all responding in the same way to the uniform, inanimate and uncreative external forces."

"But we *are* all meat machines," replied Tesla, "and it happens that I am a much more sensitive machine than other people and I receive impressions to which they are inert, and I can both understand and interpret these impressions. I am simply a finer automaton than others," he insisted.

"This difference, which you admit between yourself and others, Dr. Tesla, completely disproves your theory, from my viewpoint," I responded. "Your sensitiveness would be a purely random incident. In the integration of this randomness, with respect to all individuals, all of us would probably once, possibly very much more frequently, rise to the height of manifesting genius as you have done all your life. Even though the strokes of genius would manifest intermittently, all such individuals would receive the permanent rating as geniuses. Genius does not manifest, even intermittently, in all of us, so your meat-machine theory appears, to me, untenable. If you were really frank with me, you would tell me of many experiences you have had, strange experiences, that you could not explain, that do not fit into your meat-machine theory, and which you have been afraid to discuss with anyone for fear they would misunderstand you and perhaps ridicule you. I, however, will not find these experiences strange and beyond understanding, and one of these days you will open up and tell me about them."

As happened whenever I disagreed with him, after that evening I did not see Tesla for a while. In due time, however, I had a great many telephone conversations with him. Our discussion seemed to have brought about a change in his attitude toward me; and the next time I saw him he confided, "Mr. O'Neill, you understand me better than anyone else in the world." I mention this to indicate the correctness of my belief that there was another Tesla hidden within that synthetic individual, the superman, which Tesla sought to pass off on the public as his real self.

I did not, at this time, know about Tesla's "astounding discovery," or of some of his experiences about which I later learned. Had I known of these, my discussion with him could have been more specific.

# SEVENTEEN

ALTHOUGH Tesla thoroughly disbelieved in psychical phenomena, as previously indicated, he had many experiences which
belong in this category; and he neither discredited nor disavowed
their reality. Such paradoxes were common in all matters concerning him.

Tesla, for example, completely rejected telepathy as a phase
of psychical phenomena, but he was firmly convinced that mind
could communicate directly with mind. When asked about his
belief in telepathy by a newspaper reporter in the early nineties,
Tesla replied: "What is usually taken as evidence of the existence
of telepathy is mere coincidence. But the working of the human
mind through observation and reason interests and amazes me."
And then he added the paradoxical statement: "Suppose I make
up my mind to murder you. In an instant you would know it.
Now, isn't that wonderful enough? By what process does the
mind get at all this?"

Reduced to its simplest terms, this interview states: Psychical
telepathy does not exist as a reality; but the transmission of
thought from mind directly to mind is a wonderful phenomenon, worthy of scientific study.

The paradox here is due to the fact that, at the period in
which Tesla was speaking, all psychical phenomena were supposed to be mediated by the intervention of spirits, or souls of
the departed. Such a theory had no place in Tesla's philosophy,
since he did not believe in immortality and felt that he could explain all phenomena in terms of matter and energy; and the
spirit was supposed to lie beyond both of these categories. Thinking, however, was, according to Tesla's theories, something which
resulted from the interaction of matter and energy in the brain;
and as this process probably produced waves in the ether, there

was no reason why the waves sent out by one mind should not be received by another, with resulting transfer of thought.

Tesla would not discuss anything bordering on psychical experiences outside the circle of his relatives, however. On one occasion, though, he probably saved the lives of three of his friends through a premonition; and he related the incident to his nephew, Sava N. Kosanovich, who thus retells it:

"I heard from Tesla that he had premonitions. He explained this in a mechanical way, saying he was a sensitive receiver that registers any disturbance. He declared that each man is like an automaton which reacts to external impressions.

"He told me of one instance in which he had held a big party here in New York for some of his friends who planned to take a certain train for Philadelphia. He felt a powerful urge not to let the friends depart as planned and forcibly detained them so that they missed the train on which they had planned to travel. This train met with an accident in which there were a large number of casualties. This happened sometime in the 90's.

"When his sister Angelina was ill, and died, he sent a telegram in which he said: 'I had a vision that Angelina was arising and disappearing. I sensed all is not well.' "

Tesla himself tells a most remarkable story of two supernormal events, in an unpublished manuscript. It records a situation in which, owing to overwork, his strange phenomenon of visualization disappeared, or died, and was reborn. In coming back, it grew up quickly by repeating the visualization of events of earliest childhood and successively re-enacting later events, until it brought him to the actual moment and capped the climax by then presenting a visualization of an event that had not yet taken place.

The story of this experience, as told by Tesla:

I will tell of an extraordinary experience which may be of interest to students of psychology. I had produced a striking phenomenon with my grounded transmitter and was endeavoring to ascertain its true significance in relation to the currents propagated through the earth. It seemed a hopeless undertaking and for more than a year I worked unremittingly but in vain. This profound study so entirely absorbed me that I became forgetful of everything else, even of my undermined health. At last, as I was on the point of breaking down, nature applied the preservative, inducing lethal sleep.

Regaining my senses, I realized with consternation that I was unable to visualize scenes from my life except those of infancy, the very first ones that had entered my consciousness. Curiously enough, these appeared before my vision with startling distinctness and afforded me welcome relief. Night after night, when retiring, I would think of them and more and more of my previous existence was revealed. The image of my mother was always the principal figure in the spectacle that slowly unfolded, and a consuming desire to see her again gradually took possession of me.

This feeling grew so strong that I resolved to drop all work and satisfy my longing. But I found it too hard to break away from the laboratory and several months elapsed during which I succeeded in reviving all the impressions of my past life up to the spring of 1892.

In the next picture that came out of the mist of oblivion, I saw myself at the Hotel de la Paix in Paris just coming to from one of my peculiar sleeping spells caused by prolonged exertion of the brain. Imagine the pain and distress I felt when it flashed upon my mind that a dispatch was handed to me at that very moment bearing the sad news that my mother was dying.

It was especially remarkable that all during this period of partially obliterated memory I was fully alive to everything touching on the subject of my research. I could recall the smallest details and the least insignificant observations in my experiments and even recite pages of texts and complex mathematical formulae.

This was a prevision of the event which took place immediately after his Paris lecture, as described in an earlier chapter, in which he rushed home in time to see his mother just before she died.

The second incident also concerns the death of his mother, and is told in another connection in the same manuscript. He states:

For many years I have endeavored to solve the enigma of death and watched eagerly for every kind of spiritual indication. But only once in the course of my existence have I had an experience which, momentarily, impressed me as supernatural. It was at the time of my mother's death.

I had become completely exhausted by pain and long vigilance and one night was carried to a building about two blocks from our home. As I lay helpless there, I thought that if my mother died while I was away from her bedside she would surely give me a sign.

Two or three months before I was in London in company with my late friend, Sir William Crookes, when spiritualism was discussed and I was under full sway of these thoughts. I might not have paid attention to other men but was susceptible to his arguments as it was his epochal work on radiant matter, which I had read as a student, that made me embrace the electrical career.

I reflected that the conditions for a look into the beyond were most favorable, for my mother was a woman of genius and particularly excelling in the powers of intuition. During the whole night every fiber of my brain was strained in expectancy, but nothing happened until early in the morning when I fell into a sleep or perhaps a swoon, and saw a cloud carrying angelic figures of marvelous beauty, one of whom gazed upon me lovingly and gradually assumed the features of my mother. The apparition slowly floated across the room and vanished and I was awakened by an indescribably sweet song of many voices. In that instant a certitude, which no words can express, came upon me that my mother had just died. And that was true.

I was unable to understand the tremendous weight of the painful knowledge I received in advance and wrote a letter to Sir William Crookes while still under the domination of these impressions and in poor bodily health.

When I recovered I sought for a long time the external cause of this strange manifestation and to my great relief, I succeeded after many months of fruitless effort. I had seen the painting of a celebrated artist, representing allegorically one of the seasons in the form of a cloud with a group of angels which seem to actually float in the air, and this had struck me forcibly. It was exactly the same that appeared in my dream with the exception of my mother's likeness. The music came from the choir in the church nearby at the early mass of Easter morning, explaining everything satisfactorily in conformity to scientific facts.

This "scientific" explanation by Tesla is, of course, totally unscientific. It ignores the three principal facts: one, that he had what he identified at the time as a supernormal experience that brought with it a certitude that words could not describe; two, that this experience conveyed a revelation of his mother's death, which he understood as such; and, three, that the event took place at the exact time of her death. The mechanism by which the phenomenon was produced utilized the memories stored in Tesla's mind (of the painting, for example) as the vehicle by which the information could be presented to him in understandable, though symbolic, form. In addition, there was the premonition given several months previously as the climax of an extended phenomenon involving his mother.

Tesla's efforts to explain away "scientifically" everything of a psychical or spiritual nature, and the inadequate explanations which were satisfactory to him for this purpose, are an indication of a conflict that was taking place within him in an effort to reconcile the purely materialistic "matter and energy" superman, into which he fashioned himself, with the underlying individual into which was born a great capacity for manifesting a deep spiritual insight into life, but which he suppressed.

ONE of the strangest luncheon parties Tesla ever staged was that given by him to a prize fighter, Fritzie Zivic. It was served in one of the private dining rooms of the Hotel New Yorker in 1940. Fritzie Zivic was scheduled to take part in a prize fight at Madison Square Garden for the welterweight championship, and the luncheon was held at noon on the day of the battle.

Fritzie was one of six brothers, all of whom were either professional prize fighters or wrestlers. They lived at Pittsburgh where their father conducted a beer saloon. They were all born in Pittsburgh, but were the sons of parents, natives of Yugoslavia, whose difficult-to-pronounce Slavonic name was shortened to Zivic by the brothers for their professional activities.

Tesla had all six of the brothers as his guests. The only other guests were William L. Laurence, science writer of the New York *Times,* and the author.

*267*

Three very different types of individuals were gathered around the table. The six fighting brothers were all fine physical specimens. They averaged medium height but their powerful, chunky bodies, deep chests and broad shoulders made them seem rather short. All were clear eyed, had clear complexions and clean-cut features, were conservatively dressed in sack suits, and wore white linen collars. The two newspapermen presented an appearance in strong contrast with the fighters, and in contrast with all the others was Tesla. Laurence, with his great mop of jet black hair combed straight back, looked more like a musician.

Tesla was seated at the head of the table. At his right sat Fritzie and next to him ranged three of his brothers. Opposite them sat two other brothers and Mr. Laurence. The author sat at the far end of the table.

Tesla did not arrange one of his famous duck dinners for this occasion—he had other plans. As soon as the party was seated, Tesla stood up. The broad, stocky Fritzie looked like a pygmy by comparison. Tesla was attired in a light-weight, tight-fitting, black, single-breasted sack suit which made him look more slender than usual. He had lost considerable weight in the preceding year, and this accentuated the sharp, bony contour which his face had taken on in his latter years. His face, of the ascetic type, now was crowned with thinning locks of silvery white hair. His long slender hands, delicately shaped, started to wave over the seated prize fighter, who smiled up at the strange figure towering above him.

"I am ordering for you a nice thick beefsteak, two inches thick, so that you will have plenty of strength tonight to win the championship by a . . ."

The fighter had both hands up, trying to interrupt the gesticulating figure of the scientist.

"No," protested Fritzie, "I am in training and I cannot eat a steak today."

"You listen to me," shouted the insistent voice of Tesla, whose swinging arms and swaying body made him appear to be going through the antics of a cheer leader at a football game. "I'll tell you how to train. You will train on beefsteak. I am going to get you a beefsteak two inches thick and dripping with blood so that you will be able to . . . ."

The five brothers now joined Fritzie in his protest.

"He can't eat beeksteak today. He would lose the fight, Dr. Tesla," they chorused.

"No, he won't lose the fight," shot back Tesla. "You must think of the heroes of our national Serbian poetry. They were red-blooded men and mighty fighters. You too must fight for the glory of Serbia, and you need beefsteak dripping in blood to do it!"

Tesla had worked himself into a fine frenzy and was waving his arms and punching his palms as if he were at the ringside at an exciting moment in the battle. His frenzy was lost on Fritzie and his brother pugilists. They were unmoved. Fritzie replied:

"I will win, Dr. Tesla. I will fight for the glory of Yugoslavia and when the referee gives me the decision and I speak into the microphone I will also say I fought for Dr. Tesla—but no beef-steak today, Dr. Tesla, please."

"All right, Fritzie, you can have whatever you want," Tesla agreed, "but your brothers will have their beefsteak."

"No, Dr. Tesla," replied the eldest brother, "if Fritzie cannot have beefsteak neither will we. We will eat whatever he eats."

Fritzie ordered scrambled eggs on toast, with bacon, and a glass of milk. The five brothers gave duplicate orders and the two newspapermen did likewise.

Tesla laughed heartily. "So that is what you do your fighting on today," he said between chuckles.

For himself, the blood-thirsty 83-year-old scientist ordered "A dish of hot milk"; and on this diet he managed to summon a tremendous amount of energy during the meal which he directed toward urging Fritzie to give his opponent "everything you've got" and "make it a knockout in the first round."

It was a strange dinner. Despite the greatly outnumbering pugilists with their hard set faces and chunky powerful bodies, the thin, bony faced, sharp featured, almost emaciated scientist with his sunken eyes, and his thin, silky silver hair, easily dominated the scene. Everyone was at ease despite the brothers' anticipation of Fritzie's impending battle and Tesla's enthusiasm. Yet, in spite of the fact that everyone was relaxed, there was an eerie kind of tenseness linking the peculiar assemblage. Once I became conscious of the situation, I watched developments with interest. I had experienced such conditions previously but never under such circumstances as these.

Mr. Laurence, of the *Times,* was seated at my right. He began to act a bit restless while only halfway through the meal. Several times he looked under the table. He in turn rubbed his ankle, his knee, his calf. He shifted his position. He rubbed his elbow and later his forearm. I managed to catch his eye.

"Anything bothering you, Bill?" I asked, knowing full well what was happening.

"There is something strange going on here," he replied.

A couple of minutes later he again reached, and looked under the table.

"Feel anything?" I asked.

"Yes," he replied, seemingly a bit worried. "Something hot is touching me at different spots. I can feel the heat but I can't see anything that is doing.it. Do you feel it, too?" he asked.

"Don't worry about it," I assured him. "I know what it is, and will tell you all about it later. Just make as close observations as you can now."

The phenomenon continued until the party broke up. On our way back to our offices, I explained to Mr. Laurence.

"You have often laughed at me for my gullibility in accepting the reality of the so-called psychic experiences," I said. "Now you have had one. As soon as that luncheon got well under way, after Dr. Tesla's fiery outburst had quieted down, I sensed a peculiar tenseness in the air around me. At times the atmosphere seemed webby to my face and hands, so I suspected something unusual might happen.

"That gathering was a perfect set-up for a psychic seance, and if it was held in the dark there is no telling what we might have observed. Here were six powerfully built men, closely en rapport with each other, all filled to the bursting point with vital energy waiting for an event that would release an emotional outburst. In addition, we had Dr. Tesla staging an emotional outburst the like of which he probably never before exhibited throughout his life. He was supercharged with a different kind of vital energy. Just visualize Dr. Tesla as a medium acting as a co-ordinator, in some unknown way, to release these pent-up stores of vital energy which, again in an unknown manner, organized channels of conduction through which this energy was transferred from levels of high potential to levels of lower potential.

"In this case we were the levels of lower potential, for I had exactly the same experiences you had, with these energy-transfer channels in space making contact with various parts of my body and producing areas in which I, too, experienced a sensation of intense heat.

"You have read reports of séances in which the sitters reported that they experienced cool breezes. In these situations the action is the reverse of what we experienced, for in the séances energy was being drawn from the sitters to be organized by the so-called medium for the production of phenomena.

"Some kind of highly attenuated energy-bearing fluid was, in our experience today, drawn from the bodies of the fighters and fed into our bodies—and in the séances it is drawn from the bodies of the sitters and fed into that of the medium, or to a central collecting point. In a report which I have written on my séance observations, I have called this substance psynovial fluid, which is merely a convenient abbreviation for new psychic fluid.

"Now that you have had today's experience, you will understand why a few years ago I risked having Dr. Tesla figuratively massacre me when I told him he was using his meat-machine philosophy of human life to cover up a lot of strange experiences he has had, and about which he was afraid to talk. . . ."

ANOTHER strange supernormal experience came to Tesla a few days before he died, but he was probably totally unaware that the situation had any unusual aspects.

Early one morning he called his favorite messenger boy, Kerrigan, gave him a sealed envelope, and ordered him to deliver it as quickly as possible. It was addressed to "Mr. Samuel Clemens, 35 South Fifth Ave., New York City."

Kerrigan returned in a short time with the statement that he could not deliver the message because the address was incorrect. There is no such street as South Fifth Ave., the boy reported; and in the neighborhood of that number on Fifth Ave. no one by the name of Clemens could be located.

Tesla became annoyed. He told Kerrigan: "Mr. Clemens is a very famous author who writes under the name Mark Twain, and you should have no trouble locating him at the address I gave you. He lives there."

Kerrigan reported to the manager of his office and told him of his difficulty. The manager told him: "Of course you couldn't find South Fifth Avenue. Its name was changed to West Broadway years ago, and you won't be able to deliver a message to Mark Twain because he has been dead for twenty-five years."

Armed with this information, Kerrigan returned to Tesla, and the reception accorded his announcements left him still further confused.

"Don't you dare to tell me that Mark Twain is dead," said Tesla. "He was in my room, here, last night. He sat in that chair and talked to me for an hour. He is having financial difficulties and needs my help. So you go right back to that address and deliver that envelope—and don't come back until you have done so." (The address to which he sent the messenger was that of Tesla's first laboratory!)

Kerrigan returned to his office. The envelope, not too well sealed, was opened in the hope it would give some clue as to how the message could be delivered. The envelope contained a blank sheet of paper wrapped around twenty $5 bills! When Kerrigan tried to return the money, Tesla told him, with great annoyance, either to deliver the money or keep it.

The last two decades of Tesla's life were filled with many embarrassing situations concerning unpaid hotel bills, and it would seem that by some process of transference this situation was shifted to his perception of Mark Twain.

In view of Tesla's highly intensified abilities to see the subjects of his thoughts as materialized objects, the simpler theory would be that by his usual process he had summoned the vision of Mark Twain. Tesla and Mark Twain were very good friends, and the inventor had every reason for knowing that the heavy-thinking humorist was dead. Such being the case, how was he able to forget his death? An objective theory can be offered which may, or may not, contain the correct explanation.

Tesla's memory was filled with many recollections of Mark Twain, dating back to his early youth when he credited the reading of one of the humorist's books with having brought him out of a critical illness. Twenty years later, when Tesla related this incident, the humorist was so deeply affected he wept. A close friendship followed, filled with many pleasant incidents. Every incident concerning Mark Twain was laid down in Tesla's memory. How these records are filed in the brain we do not know, but we might assume, for the moment, that the arrangement is orderly enough, with the system based on a time sequence in which each successive incident is filed on an earlier one, the latest ones being on top. When Tesla started the process of visualizing Mark Twain in his room (and it probably operated on a subconscious level), he penetrated through the stack of memory records until he reached one that was satisfactory, and then concentrated so heavy a flow of vital energy in carrying this to the visualization center of his brain that it burned out, and destroyed, or narcotized, all later memory records that lay above it. As a result, after the visualization process was over, there was no record in Tesla's memory files of anything that happened in his relations with Mark Twain, following the pleasant record he had so strangely relived. All subsequent memory records were wiped out, including his memory of Mark Twain's death. It would then be perfectly logical for him to reach the conclusion that Mark Twain was still alive!

Several versions of this story are in circulation. They all have in common Tesla's belief that Mark Twain was still alive; that he himself had very recently been in communication with him, and sought to send him money to meet a difficult situation.

PIRATED,[*] lied about, ignored, Tesla carried on his work during the latter decades, always hoping that he would be able to arrange matters so that he would be able to finance all the inventions he was treasuring in his mind. His pride would not permit him to admit financial embarrassment. He was forced frequently to leave hotels because of unpaid bills. His friend, B. A. Behrend, author of the book, *The Induction Motor*, which had clarified Tesla's theory for engineers, when visiting New York and finding the inventor moved from the hotel where he last found him, in each instance paid his bills, and caused Tesla's held baggage to be forwarded to him.

In the early thirties, when it seemed as if financial discouragements would "have him down," Tesla, however, appeared as optimistic as ever. He declared: "It is impossible for anyone to gain any idea of the inspiration I gain from my applied inventions which have become a matter of history, and of the force it supplies to urge me forward to greater achievements. I continually experience an inexpressible satisfaction from the knowl-

* Dr. W. H. Eccles concludes an obituary memorial, in *Nature* (London), February 13, 1943:—"Throughout his long life of 85 years Tesla seldom directed attention to his own successes, never wrote up again his old work, and rarely claimed priority though continually pirated. Such reserve is particularly striking in a mind so rich in creative thought, so competent in practical achievement."

edge that my polyphase system is used throughout the world to lighten the burdens of mankind and increase comfort and happiness, and that my wireless system, in all of its essential features, is employed to render a service to and bring pleasure to people in all parts of the earth."

When his wireless-power system was mentioned, he exhibited no sign of resentment over the collapse of his project but replied philosophically: "Perhaps I was a little premature. We can get along without it as long as my polyphase system continues to meet our needs. Just as soon as the need arises, however, I have the system ready to be used with complete success."

On his eightieth birthday he was asked if he expected actually to construct and operate his recently announced inventions, and in reply he quoted, in German, a stanza from Goethe's *Faust:*

> The God that in my bosom lives
> Can move my deepest inmost soul,
> Power to all my thought he gives
> But outside he has no control.

It had been Tesla's intention to write his autobiography. He desired to have the story of his work recorded with a most meticulous regard for accuracy; and this, he felt, no one but himself could bring to it. He declared that he had no intention of starting work on this project until he had accomplished the application of all of his major discoveries. Several persons who proposed writing his biography received only a refusal of the requested co-operation. Kenneth Swezey, a writer on scientific subjects, maintained close contact with Tesla for a number of years, and it was expected that Tesla would co-operate with him in such a project. Swezey assembled seventy letters from leading scientists and engineers in all parts of the world as a surprise for Tesla on his seventy-fifth birthday, at which time the letters, bound in a memorial volume, were presented to him. These letters, reprinted in Yugoslavia, led to the establishment of the Tesla Institute in that country. Swezey was engaged in war work and expected, at the time of Tesla's death, to enter the Navy; otherwise he might have undertaken the task of writing Tesla's biography. Tesla, even up to his eighty-fourth year, expected to recover more robust health and to live beyond the century mark. It is probable, therefore, that he had not started work on his auto-biography. Whether or not any parts of it have been written is impossible to ascertain at the present time. All Tesla's records were sealed by the Custodian of Alien Property, although Tesla was a citizen of the United States.

During the last half-dozen years of his life, Tesla, happily, was supplied with enough money to meet his immediate needs, thanks to the payment to him of an honorarium of $7,200 a year, by the Yugoslav government, as patron of the Tesla Institute, established in Belgrade.* Even with this income, however, and with a very limited range of activity (being confined largely to his room), during the last two years Tesla still managed to fall behind in his hotel bill. This was owing to his unlimited generosity. He was very generous in bestowing tips on all who performed the slightest service for him, and in responding to the slightest suggestion that anyone was in need of assistance.

During the latter part of 1942 he spent most of the time in bed, mentally active but physically weak. He permitted no visitors to come to his room, not even close associates of earlier years. He insisted to hotel employees that he was not ill and refused to listen to suggestions that he have a doctor visit him. He gave orders that even hotel employees were not to enter his room unless he summoned them.

On January 5, Tuesday morning, he permitted the maid to

---

* The Society for the Foundation of the Tesla Institute at Belgrade was organized as Tesla neared his eightieth year. It enlisted support from the scholars, the government, commercial interests and the people as a whole. From the government and private sources an endowment was subscribed which was adequate to erect and equip a research laboratory and maintain it in operation as an institute. The Institute was opened in 1936, in commemoration of Tesla's eightieth anniversary. A week of observance was held throughout Yugoslavia and formal celebrations were held at Belgrade on May 26, 27 and 28, at Zagreb on May 30, and at his native village, Smiljan, on June 2 and also on July 12.

come to his room, and then gave orders to guard his room closely so that he would not be disturbed. This was done. It was not unusual for Tesla to give orders that he was not to be disturbed for protracted periods. Early Friday morning (January 8) a maid with a premonition, risking his displeasure, entered Tesla's room and found him dead. He looked peaceful, as if resting, with a suggestion of a smile on his gaunt bony face. The superman died as he had lived—alone.

THE police were notified that Tesla had died alone and without medical attendance. The coroner declared his death due to natural causes incident to senility; and that he had died on the night of Thursday, January 7, 1943, some hours before the maid entered the room. Operatives from the Federal Bureau of Investigation came and opened the safe in his room and took the papers it contained, to examine them for a reported important secret invention of possible use in the war. The body was removed to Campbell's Funeral Parlors at Madison Avenue and 81st Street.

Funeral services were held at the Cathedral of St. John the Divine, on Tuesday, January 12, at 4 P.M. Bishop Manning offered the opening sentences of the Burial Office and Final Prayer. Following the services, the body was removed to Ferncliff Cemetery at Ardsley, N. Y., and was later cremated.

# FIFTH PART

## *AFTERGLOW*

# EIGHTEEN

DESPITE his celibate life, and his almost hermitlike existence in his own intellectual sphere, Tesla was, in his social contacts, a charming individual. The year he had spent digging ditches and doing hard manual labor, when he could get a job of any kind, and his experience during that time of sleeping in any shelter he could obtain and eating any kind of food he could manage to secure, undoubtedly made a tremendous and lasting impression on him. The fact that he could never be induced to discuss this period would so indicate. Yet it probably softened him in a beneficial way—by a going-through-the-mill process. But it had been a grievous insult to his personality to be valued only for the brute strength in his muscles; and this rankled ever after.

Once he had obtained funds through the founding of his laboratory and the sale of his patents to Westinghouse, he thereafter maintained an almost princely status. He knew how to wear clothes to increase the impressiveness of his appearance; his tallness gave him something of an advantage over others; his obvious physical strength brought him a respect that forbade any invasion of his attitude; his excellent English and the care he exercised to use the language correctly, and his command of a half-dozen other languages, established him as a scholar; and the first batch of his alternating-current inventions created for him in the mind of the public a reputation for outstanding scientific accomplishment. The fact that he always spoke of the value of his inventions to the world, and not of the greatness of his own accomplishment, endeared him to all who met him.

When Tesla was riding a tidal wave of popularity during the nineties, he was averse to publicity; but frequently well-known writers for the newspapers were able to break through the barriers and secure "feature" articles. An excellent description of him, keyed to the manner of the period, is contained in an article written by Franklin Chester, in the *Citizen* of August 22, 1897. The portion referring to his personal appearance and activities follows:

So far as personal appearance goes no one can look upon him without feeling his force. He is more than six feet tall and very slender. Yet he possesses great physical power. His hands are large, his thumbs abnormally long, and this is a sign of great intelligence. His hair is black and straight, a deep shining black. He brushes it sharply from over his ears, so that it makes a ridge with serrated edges.

His cheekbones are high and prominent, the mark of the Slav: his skin is like marble that age has given the first searing of yellow. His eyes are blue, deeply set, and they burn like balls of fire. Those weird flashes of light he makes with his instruments seem also to shoot from them. His head is wedge shaped. His chin is almost a point.

Never was a human being filled with loftier ideals. Never did a man labor so unceasingly, so earnestly, so unselfishly for the benefit of the race. Tesla is not rich. He does not trouble himself about money. Had he chosen to follow in the footsteps of Edison he could be, perhaps, the richest man in the world, and Tesla is just 40 years old.

Tesla is, above all things, a serious man, undoubtedly the most serious man in New York. Yet he has a keen sense of humor and the most beautiful manners. He is the most genuinely modest of men. He knows no jealousy. He has never decried the accomplishments of another, never refused credit.

When he talks you listen. You do not know what he is saying, but it enthralls you. You feel the importance without understanding the meaning. He speaks the perfect English of a highly educated foreigner, without accent and with precision. He speaks eight languages equally well.

The daily life of this man has been the same, practically, ever since he has been in New York. He lives in the Gerlach, a very quiet family hotel, in 27th street, between Broadway and Sixth avenue. He starts for his laboratory before 9 o'clock in the morning, all day long he lives in his weird, uncanny world, reaching forth to capture new power to gain fresh knowledge.

No stranger ever sees him at his work. No one knows of his assistants. At rare intervals he presents some experiments in his laboratory, and there is no sacrifice that thousands of people would not make to gain admission to these.

Usually he works until 6 o'clock, but he may stay later. The absence of natural light does not trouble him. Tesla makes sunlight in his workshop.

At exactly 8 o'clock he enters the Waldorf. He is attired in irreproachable evening clothes. In the winter time he never wears an evening jacket, but always the coat with tails.

He finishes his dinner at exactly 10 o'clock, and leaves his hotel, either to go to his rooms to study or to return to his laboratory to work through the night.

Arthur Brisbane, who later became Hearst's famous editor, interviewed Tesla and published in *The World,* August 22, 1894, the longest story he had written on a famous person. He declared Tesla "Our Foremost Electrician—Greater Even than Edison," and included the following description of him:

He has eyes set very far back in his head. They are rather light. I asked him how he could have such light eyes and be a Slav. He told me that his eyes were once much darker, but that using his mind a great deal had made them many shades lighter. I have often heard it said that using the brain makes the eyes lighter in color. Tesla's confirmation of the theory through his personal experience is important.

He is very thin, is more than six feet tall, and weighs less than a hundred and forty pounds. He has very big hands. His thumbs are remarkably big, even for such big hands. They are extraordinarily big. This is a good sign. The thumb is the intellectual part of the hand. The apes have very small thumbs. Study them and you will notice this.

Nikola Tesla has a head that spreads out at the top like a fan. His head is shaped like a wedge. His chin is as pointed as an ice-pick. His mouth is too small. His chin, though not weak, is not strong enough.

*283*

His face cannot be studied and judged like the faces of other men, for he is not a worker in practical fields. He lives his life up in the top of his head, where ideas are born, and up there he has plenty of room. His hair is jet black and curly. He stoops—most men do when they have no peacock blood in them. He lives inside of himself. He takes a profound interest in his own work. He has that supply of self-love and self-confidence which usually goes with success. And he differs from most of the men who are written and talked about in the fact that he has something to tell.

Tesla had, to be sure, a sense of humor and enjoyed putting over a subtle joke. Before the period in which he became a regular diner at the Waldorf-Astoria, he dined nightly at Delmonico's, then the smartest hostelry in the city, and a gathering place for "The 400." Tesla was the most famous and spectacular figure among the famous patrons of the famous place, but he always dined alone. He could never be induced to join other groups and never had a guest of his own. After dining he would always return to work at his laboratory.

One evening some of his friends, believing that he was working too hard and should get some relaxation, induced him to join them in a game of billiards. They assumed he had neglected to learn how to play games, so, on arriving at the billiard room, they explained to him how to hold the cue, strike the balls, and other elements of the game. Tesla had not played billiards in a dozen years; but during his second year at Gratz, when he was a year ahead in his studies and spent his evenings in the cafés, he had become an expert billiardist. When the experts at Delmonico's gave him preparatory instruction, he asked some "dumb" questions, and made some intentional miscues. Taking on one of the players and still asking silly questions, he tried the most difficult way of making shots—to demonstrate his purely amateur status—and made them, to the amazement of the experts. Several of them took him on that evening, and he defeated all of them with badly unbalanced scores. He declared the new game give him a wonderful opportunity to practice very abstract mathematical theories; and the experts at Delmonico's spread stories about the wonderful accomplishment of Scientist Tesla in mastering the game in a single evening and defeating the best players in the city. The story got into the newspapers. Tesla refused to play any more, declaring he was in danger of becoming so enthusiastic over the game that it would interfere with his researches.

This same man magnificent who graced the Waldorf-Astoria and Delmonico's was not averse, however, to visiting the Bowery, which was but a block away from his Houston Street laboratory. He repaired to a thirst-quenching emporium on that thoroughfare one afternoon shortly after a denizen of the Bowery, Steve Brodie, had achieved fame by jumping, or at least claiming to have jumped, off the Brooklyn Bridge. As Tesla raised his glass of whiskey he said to the bartender: "You know what Steve said as he was about to jump off the bridge—'Down he goes' "; and with that he downed his liquor in a gulp.

A near-by drinker, a little the worse for several, misunderstood Tesla's remark and got the impression he had heard Steve Brodie telling the final episode of his feat. He rushed up to Tesla to buy him a drink, and was joined by his friends. Tesla with a laugh shook them off and dashed out of the bar, while the misguided drinker started after him yelling, "Stop him, that's Steve." On the street the pedestrians misunderstood the thick-tongued drinker's shout and joined him in the chase, calling "Stop, thief!" Tesla's long legs rendered him a valuable service and he got a lead on the crowd, dashed into an alley, over a fence and climbed a fire escape on the back of his own building, reached his laboratory through a window, quickly donned a blacksmith's apron and started hammering a bar of metal. His pursuers, however, failed to trace him.

Tesla was idolized by the Serbians in New York. A great many of them could claim to be distant relatives through either the Tesla or Mandich side of the family, and those who could not claim this distinction revered him none the less, despite the fact he never accepted invitations to take part in their social or other functions.

One day an excited Serbian, a laborer, came to his apartment at the Waldorf-Astoria to beg his aid. He had gotten into a fight and pummeled a fellow Serbian, who had sworn out a warrant for his arrest. The visitor did not have any money but wanted to go to Chicago to escape arrest. Would Tesla please lend him the money for his railroad fare?

"So you assaulted a man and now want to run away to escape punishment," said Tesla. "You may run away from the law but you are not going to escape punishment; you are going to get it right now!" Seizing a cane and grasping the man by the back of the neck, he ran him around the room, beating the dust out of the seat of his trousers until the man cried for mercy.

"Do you think you can be a better man in Chicago and keep out of fights?" Tesla asked him. The man was sure he could. He received the money for his railroad fare and a few dollars more.

So great was Tesla's popularity in the nineties that many persons came to dine in the Palm Room at the Waldorf just to catch a glimpse of the famous inventor. He arranged to leave his office at six, but just before leaving he would telephone the order for his dinner to the headwaiter, always insisting that none less could serve him. The meal was required to be ready at eight o'clock. In the meantime he would go to his room and array himself in formal evening attire—white tie and tails. He dined alone, except on the rare occasions when he would give a dinner to a group to meet his social obligations.

Money was always a nuisance detail to Tesla. For about fifteen years, following 1888, he always had all he needed to meet his obligations; and he lived well. After about 1902 his financial road became quite rocky—but his fame was greater than ever, and likewise the need for maintaining his standard of living if he was to recoup his fortune. He continued to stage frequent large dinners at the Waldorf to repay his social obligations, and had difficulty in accustoming himself to a money deficiency. On one occasion, when a large party was assembled in a private dining room, the headwaiter whispered to him that a most excellent dinner was prepared and ready to serve as he had ordered it, but that the credit department insisted it could not be served until he paid for it in advance. "Get Mr. Morgan on the telephone in the manager's office and I will be down there immediately," Tesla fumed. In a short time a more-than-adequate check was delivered to Tesla by a messenger. Many such occasions are reported to have arisen, but were always straightened out in the manager's office, usually without any outside intervention.

The closest approach to home life which Tesla enjoyed came to him through Robert Underwood Johnson, diplomat and poet, and one of the editors of the *Century Magazine,* whose home was in Madison Avenue in the fashionable Murray Hill district. Tesla and Johnson were very close friends. A love of poetry was one of the several interests they had in common. Johnson wrote, and published in the *Century,* in April, 1895, a short poem on his visit to Tesla's laboratory. This led to a co-operative enterprise in which he paraphrased many pieces of Serbian poetry from literal translations made by Tesla, who could recite many thousands of lines of such material from memory. About forty pages of these translations, with an introductory note by Tesla, appeared in the next edition of *Poems* by Johnson.

Persons famous in all fields of activity were frequent guests in the Johnson home, and formal dinners were constantly being held for brilliant assemblages of personalities. Tesla was present as frequently as he could be induced to come, but he preferred to avoid all formal dinners as much as possible. He was, however, a very frequent informal visitor, arriving unexpected, and often at most unusual hours. It was not uncommon for Tesla to arrive at the Johnson home after midnight, after the family had retired, and for "Bob" and "Nick" to sit up for hours reveling in the exchange of a magnificent array of ideas. (Johnson and "Willie" K. Vanderbilt, were, as has been noted, the only individuals who rated the exchange of first names with Tesla.)

Tesla's visits to the Johnson home were always many hours long. He would arrive in a hansom cab, which he always required to wait for him to return to his hotel only a few blocks distant. The Johnson children learned to take advantage of this, and when he arrived early in the evening they would get his permission to use the cab for a drive through Central Park while he chatted at home.

Tesla enjoyed the opera and at one time attended the performances quite frequently. William K. Vanderbilt's box was always available to him, as likewise were those of many other patrons of the Metropolitan. He occasionally attended the theatre. His favorite actress was Elsie Ferguson who, he declared, knew how to dress and was the most graceful woman he had ever seen on the stage. He gradually dropped both the theatre and opera in favor of the movies, but was an infrequent attendant even at those. He would not witness a tragedy but enjoyed comedy and the lighter aspects of entertainment.

One of his close friends was Rear Admiral Richmond Pearson Hobson, the Spanish American War hero. In later years, Hobson was the only person who was able to cajole Tesla into breaking a long vigil at his intellectual pursuits for a session at the movies.

Tesla did not subscribe to any religion. Early in life he severed his relations with the Church and did not accept its doctrines. At his seventy-fifth birthday dinner he declared that that which is called the soul is merely one of the functions of the body, and that when the activities of the body cease, the soul ceases to exist.

IT IS difficult for a man to appear as a hero to his secretary, but to Miss Dorothy F. Skerritt, who served Tesla in this capacity for many years until he closed his office when he was seventy, he remained a saintly superman. Her description of Tesla, at this age, records him as possessing the same magnetic personality that so impressed writers thirty years earlier. She wrote:

As one approached Mr. Tesla he beheld a tall, gaunt man. He appeared to be an almost divine being. When about 70 he stood erect, his extremely thin body immaculately and simply attired in clothing of a subdued coloring. Neither scarf pin nor ring adorned him. His bushy black hair was parted in the middle and brushed back briskly from his high broad forehead, deeply lined by his close concentration on scientific problems that stimulated and fascinated him. From under protruding eyebrows his deepset, steel gray, soft, yet piercing eyes, seemed to read your innermost thoughts. As he waxed enthusiastic about fields to conquer and achievements to attain his face glowed with almost ethereal radiance, and his listeners were transported from the commonplaces of today to imaginative realms of the future. His genial smile and nobility of bearing always denoted the gentlemanly characteristics that were so ingrained in his soul.

Until the last, Tesla was meticulously careful about his clothes. He knew how to dress well and did so. He declared to a secretary, in 1910, that he was the best-dressed man on Fifth Avenue and intended to maintain that standard. This was not because of personal vanity. Neatness and fastidiousness in clothes were entirely in harmony with every other phase of his personality. He did not maintain a large wardrobe and he wore no jewelry of any kind. Good clothes fitted in very nicely with his courtly bearing. He observed, however, that in the matter of clothes the world takes a man at his own valuation, as expressed in his appearance, and frequently eases his way to his objective through small courtesies not extended to less prepossessing individuals.

He was partial to the waisted coat. No matter what he wore, however, it carried an air of quiet elegance. The only type of hat he wore was the black derby. He carried a cane and wore, usually, gray suede gloves.

Tesla paid $2.50 a pair for his gloves, wore them for a week and then discarded them even though they still appeared as fresh as when they came from the maker. He standardized his style of ties and always wore the four-in-hand. The design motive was of minor importance but the colors were limited to a combination of red and black. He purchased a new tie every week, paying always one dollar.

Silk shirts, plain white, were the only kind Tesla would wear. As with other articles of his clothing, such as pajamas, his initials were always embroidered on the left chest.

Handkerchiefs he purchased in large numbers because he never sent them to the laundry. After their first use they were discarded. He liked a good quality of linen and purchased a standard package brand. His collars were never laundered, either. He never wore one more than once.

Tesla always wore high-laced shoes, except on formal occasions. He required a long narrow shoe and insisted on a last that had a neatly tapered square-toe effect. His shoes were undoubtedly made to order, for the tops extended halfway up his calf, a style that could not be purchased in merchant shoe stores. His tallness in all probability made this additional support at the ankles desirable.

The single use of articles, such as handkerchiefs and collars, extended to napkins. Tesla had a germ phobia, and it acted like so much sand in the social machinery of his life. He required that the table he used in the dining room of his hotel be not used by others. A fresh table cloth was required for every meal. He also required that a stack of two dozen napkins be placed on the left side of the table. As each item of silverware and each dish was brought to him—and he required that they be sterilized by heat before leaving the kitchen—he would pick each one up, interposing a napkin between his hand and the utensil, and use another napkin to clean it. He could then drop both napkins on the floor. Even for a simple meal, he usually ran through the full stock of napkins. Flies were his pet abomination. A fly alighting on his table was adequate cause for removing everything from the table and making an entirely new start with the meal.

Tesla was fortunate in that the headwaiter at the Waldorf-Astoria, during the period he was living there, Mr. Peterson, was afterward headwaiter at the Hotel Pennsylvania, where he later lived for several years. A story was in circulation to the effect that both at the Waldorf and at the Pennsylvania a special chef was employed to prepare Tesla's meals, but Mr. Peterson states that this story was untrue.

In his earlier years, for dinner, he greatly enjoyed fine thick steaks, preferably the filet mignon, and it was not unusual for him to consume two or three at a sitting. Later his preference turned to lamb, and he would frequently order a roast saddle of it. While the saddle was usually large enough to serve a party of several persons, as a rule he ate of it only the central portion of the tenderloin. A crown of baby lamb chops was another favorite dish. He also relished roast squab with nut stuffing. In fowl, however, his choice was roast duck. He required that it be roasted under a smothering of celery stalks. This method of preparing the duck was of his own devising. He very often made it the central motif around which a dinner was designed when entertaining friends, and on such occasions he would go to the kitchen to superintend its preparation. Duck so prepared was nevertheless delicious. Of the duck he ate only the meat on either side of the breast bone.

With the passing decades, Tesla shifted away from a meat diet. He substituted fish, always boiled, and finally eliminated the meat entirely. He later almost entirely eliminated the fish and lived on a vegetarian diet. Milk was his main standby, and toward the end of his life it was the principal item of diet, served warm.

As a youth he drank a great deal of coffee, and, while he gradually became aware that he suffered unfavorable influences from it, he found it a difficult habit to break. When he finally made the decision to drink no more of it, he adhered to his good intentions but was forced to recognize the fact that the desire for it remained. He combated this by ordering with each meal a pot of his favorite coffee, and having a cup of it poured so that he would get the aroma. It required ten years for the aroma of the coffee to transform itself into a nuisance so that he felt secure in no longer having it served. Tea and cocoa he also considered injurious.

He was a heavy smoker in his youth, mostly of cigars. A sister who seemed fatally ill, when he was in his early twenties, said she would try to get better if he would give up smoking. He did so immediately. His sister recovered, and he never smoked again.

Tesla drank whiskey, for this he considered a very beneficial source of energy and an invaluable means for prolonging life. It was responsible, he believed, for the longevity enjoyed by many of his ancestors. It would enable him, he declared early in the century, to live to one hundred and fifty. When prohibition came along with the First World War, he denounced it as an intolerable interference with the rights of citizens. Nevertheless, he promptly gave up the use of whiskey and all other beverages except milk and water. He declared, however, that the elimination of whiskey would reduce his expectation of life to one hundred and thirty years.

Stimulants were not necessary to help him to think, Tesla said. A brisk walk he found much better as an aid for concentration. He seemed to be in a dream when walking. Even one whom he knew very well he would pass at close range and not see, though he might appear to be looking directly at him. His thoughts were usually miles away from where he was. It was this practice, apparently, which was responsible for the accident, in 1937, when he was struck and severely injured by a taxicab. As a matter of fact, he had stated in an interview two years earlier that he would probably be killed by a truck or taxicab while jaywalking.

Tesla's weight, stripped, was 142 pounds, and, except during brief periods of illness, hardly varied a pound from 1888 to about 1926, when he intentionally reduced his weight five pounds.

One of Tesla's indulgences, over many years, was scalp massages. He would visit a barbershop three times a week and have the barber rub his scalp for half an hour. He was insistent upon the barber placing a clean towel on his chair but, strangely enough, he did not object to the use of the common shaving mug and brush.

Tesla always claimed that he never slept more than two hours a night. His retiring time, he said, was five A.M., and he would arise at ten A.M. after spending only two hours in sleep, three hours being too much. Once a year, he admitted, he would sleep for five hours—and that would result in building up a tremendous reserve of energy. He never stopped working, he claimed, even when asleep. Tesla laughed at Edison's claim that he slept only four hours a night. It was a regular practice with Edison, he said, to sit down in his laboratory and doze off into a three-hour nap about twice a day. It is possible that Tesla, too, obtained some sleep in a similar fashion, perhaps without being conscious of the fact. Hotel employees have related that it was quite common to see Tesla standing transfixed in his room for hours at a time, so oblivious to his surroundings that they were able to work around his room without his being, apparently, aware of their presence.

Tesla always provided his office with a separate washroom which no one but himself was permitted to use. He would wash his hands on the slightest pretext. When he did so, he required that his secretary hand him a freshly laundered towel each time to dry them.

He went to extremes to avoid shaking hands. He usually placed his hands behind his back when anyone approached who he feared might make an effort to shake hands, and this frequently led to embarrassing moments. If by chance a visitor to his office should catch him off guard and shake his hand, Tesla was so upset that he would be unable to pay attention to the visitor's mission and frequently would dismiss him before it was completely stated; and immediately he would rush to the washroom and scour his hands. Workmen eating their lunch with dirty hands almost nauseated him.

Pearls, too, were one of Tesla's phobias. If a woman guest at a dinner party to which he was invited wore pearls, he was unable to eat. Smooth round surfaces, in general, were an abomination to him; it had even taken him a long time to learn to tolerate billiard balls.

Tesla never knew the experience of having a headache. In spite of a number of cases of serious illness, in his independent years he was never attended by a doctor.

There were reasons for practically all of Tesla's phobias, not all of them generally known. His germ phobia can be traced back to his two serious illnesses early in life, both of which were probably cholera, a disease constantly prevalent in his native land, caused by a germ transmitted by impure drinking water and by contact between individuals.

Tesla was not oblivious of his idiosyncrasies; he was quite aware of them and of the friction which they caused in his daily life. They were an essential part of him, however, and he could no more have dispensed with them than he could his right arm. They were probably one of the consequences of his solitary mode of life or, possibly, a contributing cause of it.

## *NINETEEN*

---

Tesla's mind always seemed to be under an explosive pressure. An avalanche of ideas was forever straining for release. He seemed to be unable to keep up with the flood of his own thoughts. He never had sufficient facilities to keep his accomplishments equal to his projects. If he had an army of adequately trained assistants, he would still be insufficiently equipped. As a result, those associated with him always experienced a sense of "drive"; yet he was a most generous employer both in the matter of wages paid and the number of hours of work required. He frequently demanded overtime work but always paid generously for it.

Nevertheless, Tesla was not an easy man to work for. He was most meticulously neat in his personal affairs and required all workers to be the same. He was an excellent mechanic and set extremely high standards, by his own accomplishments, for all work done in his shops. He greatly admired cleverness in his assistants, frequently rewarding them with extra compensation for difficult jobs well done, but was extremely impatient with stupidity and carelessness.

Although Tesla maintained a staff of draughtsmen, he never used them in his own design work on machines, and tolerated them only because of unavoidable contacts with other organizations. When having machines constructed for his own use, he would give individual instruction on each part. The workman scheduled to do the machinework would be summoned to Tesla's desk, where the inventor would make an almost microscopically small sketch in the middle of a large sheet of paper. No matter how detailed the piece of work, or its size, the sketch was always less than one inch in its largest dimensions. If Tesla made the slightest slip of the pencil in drawing the sketch, he would not make an erasure but would start over on another sheet of paper. All dimensions were given verbally. When the drawing was finished, the workman was not permitted to take it with him to the shop to guide him in his work. Tesla would destroy the drawing and require the machinist to work from memory. Tesla depended entirely on his memory for all details, he never reduced his mentally completed plans to paper for guidance in construction—and he believed others could achieve this ability if they would make sufficient effort. So he sought to force them to try by insisting on their working without drawings.

All those who worked with Tesla greatly admired him for his remarkable ability to keep track of a vast number of finest details concerning every phase of the many projects he had under way simultaneously. No employee was ever given any more information than was absolutely essential for completing a project. No one was ever told the purposes for which a machine or article was to be used. Tesla claimed that Edison received more ideas from his associates than he contributed, so he himself bent over backward to avoid this situation. He felt that he was the richest man in the world in the matter of ideas and needed none from anyone else; and he intended to prevent all from contributing any.

Tesla was probably very unfair to Edison in this respect. The two men were entirely different and distinct types. Tesla was totally lacking in the university type of mind; that is, the mind which is adapted to co-operate with others in acquiring knowledge and conducting research. He could neither give nor receive, but was entirely adequate to his own requirements. Edison had more of the co-operative, or executive, type of mind. He was able to attract brilliant associates and to delegate to them major portions of his inventive research projects. He had the ability to act as a catalyzer, to stimulate them to creative mental activities, and thus multiply his own creative abilities. If Tesla had possessed this ability, his record of accomplishment would have been tremendously magnified.

The inability to work with others, the inability to share his plans, was the greatest handicap from which Tesla suffered. It completely isolated him from the rest of the intellectual structure of his time and caused the world to lose a vast amount of creative thought which he was unable to translate into complete inventions. It is a duty of a master to train pupils who will carry on after him—but Tesla refused to accept this responsibility. Had Tesla, in his most active period, associated with him a half-dozen brilliant young scientists, they would have been in a position to link him with the engineering and scientific worlds from which, despite his eminence and his outstanding accomplishments, he was to a great extent isolated because of his unusual personal characteristics. His fame was so secure that the success of his assistants could not have detracted from it; but the master would have shone more brightly in the brilliant accomplishments of his pupils. He might well have attracted some practical young men who could have aided him by assuming the burden of making practical application of some of the minor but important inventions from which he could have earned sufficient profit to pay the cost of maintaining his laboratories. Many scores of important inventions have undoubtedly been lost to the world because of Tesla's intellectual hermit characteristics. Undoubtedly, he indirectly inspired many young men to become inventors.

Tesla responded powerfully to personal idiosyncrasies in individuals with whom he worked. When his reaction was unfavorable, he was unable to tolerate the presence of the person within eyeshot. When carrying on his experimental work at the Allis Chalmers plant in Milwaukee, for example, he did not increase his popularity by insisting that certain workers be dropped from the crew working on the turbine because he did not like their looks. Since, as noted earlier, he had already antagonized the engineers in that plant by going over their heads to the president and board of directors, the turbine job went forward in something less than a co-operative atmosphere.

Tesla was thoroughly impractical throughout, too, in handling money matters. When he was working on the Union Sulphur Company turbine project, a ship was made available for his use, free, during the day; but if he worked after six P.M. it would cost him $20 per hour. He never showed up at the ship until six o'clock. Every night, in addition, he had to hand out $10 for suppers for the crew. In the course of a year these costs totaled about $12,000, which must have cut heavily into the retainer he received. Nor were these his only additional expenses. Almost every night he handed a $5 tip to his principal assistants among the crew, and once a week to all members of the crew. These manifestations of generosity were not, of course, a total loss to Tesla; they might rather be classed as necessities, for he was very dictatorial in directing his assistants.

Inquiries among the employees at hotels where he lived revealed that he had a reputation for acting in a most cavalier manner toward the servants. He was almost cruel in the manner in which he ordered them around, but would make immediate compensation by the generous tips he bestowed.

He was always, however, very considerate of women, and even men, on his office staff. If any one of them did an unusually fine piece of work, everyone on the staff was informed of it. Criticism was always delivered privately to the individual involved.

Tesla had a standing rule that every messenger boy who came to his office was to receive a tip of twenty-five cents, and he set aside a fund of $10 a week for this purpose.

If necessity required that he keep his staff of young women secretaries and typists working overtime for several hours, he would provide them with a dinner at Delmonico's. He would hire a cab for the girls and would follow them in another cab. After making arrangements to pay the bill, and paying the tip in advance, he would leave.

Tesla timed his arrival at the office so that he entered at the stroke of noon. He required that his secretary should be standing immediately inside the door to receive him and take his hat, cane and gloves. His offices were opened by nine o'clock each morning, so all routine matters would be handled before his arrival. Before Tesla arrived, all the shades in the office had to be drawn so that no outdoor light was admitted and night conditions were simulated. The inventor, as remarked, was a "sun dodger." He appeared to be at his best at night and at some kind of disadvantage in daylight; at any rate, he preferred the night for work and what he called his recreation.

The only time Tesla would permit the shades of his office to be raised was when a lightning storm was raging. The various offices he leased faced on open spaces. The 8 West 40th Street office was on the south side of Bryant Park, in the east end of which was the low-roofed structure that housed the New York Public Library. From his windows on the twentieth floor, he could look beyond the city roof scape below him and obtain a broad view of the sky.

When the rumbles of distant thunder announced that the fireworks of the sky would presently be flashing, it was not only permissible to raise the shades—it was obligatory. Tesla loved to watch lightning flash. The black mohair couch would be drawn close to the windows so that he could lie on it, completely relaxed, while his vision commanded a full view of the northern or the western sky. He was always talking to himself, but during a lightning storm he would become eloquent. His conversation on such occasions was never recorded. He wished to be a lone observer of this gorgeous spectacle, and his secretaries were quite willing that he should be so accommodated. By finger measurements and counting seconds he was able to calculate the distance, length and voltage of each flash.

How thrilled Tesla must have been by these tremendous sparks, many times longer than he had been able to produce in his Colorado Springs laboratory! He had successfully imitated Nature's electrical fireworks, but he had not as yet exceeded her performance.

The ancient Romans sublimated their frustrations by the forces of Nature by creating the mental concept of their mightiest god, Jupiter, as one endowed with the power of creating lightning and hurling his bolts at earth. Tesla had refused to accept frustration; but, like the ancient Romans, he too set up a mental concept, a superman not inferior to the Romans' ruling god, who would control the forces of Nature. Yes, Tesla thoroughly enjoyed a lightning storm. From his mohair couch, he used to applaud the lightning; he approved of it. He may even have been a little bit jealous.

TESLA never married; no woman, with the exception of his mother and his sisters, ever shared the smallest fraction of his life. He idolized his mother and admired his sisters for their intellectual accomplishments. One of his sisters, Marica, exhibited unusual ability as a mathematician and had greater ability than his own for memorizing long passages from books. He attributed to his mother most of his abilities as an inventor, and he continuously spoke in praise of her ability to contrive useful gadgets for the household, often regretting that she had not been born into an environment in which she would have been able to manifest to a larger world her many creative talents. He was not unaware of the values which a woman could bring into a man's life, for he had ever before him the vast contributions which his mother made to his father's welfare and happiness. However, he lived instead a blueprint life, one which he had planned in his early youth, one designed along engineering lines, with all of the time and energies available to be directed to invention and none to be dissipated on emotional projects.

From the romantic point of view, Tesla as a young man was not unattractive. He was too tall and slender to pose as the physical Adonis, but his other qualifications more than compensated for this possible defect. He was handsome of face had a magnetic personality, but was quiet, almost shy; he was soft spoken, well educated and wore clothes well in spite of inadequate funds with which to keep up a wardrobe. However, he avoided romantic encounters, or any situations that would lead up to them, just as assiduously as other young men sought them. He would not permit his thoughts to wander into romantic channels, and with thoughts successfully controlled, action control became a problem of vanishing magnitude. He did not develop an antagonism to women; he solved the problem, instead, by idealizing them.

A typical instance of how he avoided romance is furnished by an incident that occurred in Paris when he returned to that city to give a lecture on his alternating-current system after he had become world famous. His wonderful discoveries were the principal topic of conversation of the day, and he was the cynosure of all eyes wherever he went. The situation was entirely pleasant to Tesla. Less than ten years before, the executives the Continental Edison Company, in that city, had not alone rejected the alternating-current system he had offered them but had cheated him of his just earnings. Now he was returning to that city after receiving recognition and wealth in the United States and fame throughout the world. He was in Paris as a returned hero and the world was at his feet.

As he sat in an outdoor café with a young male friend, amidst a chattering, fashionably dressed crowd, a graceful, gorgeously gowned young woman, with a stylishly coiffured crown of red hair, whom he instantly recognized as Sarah Bernhardt, the famous French actress—the "divine Sarah"—swung close to his table and when a few feet away very auspiciously dropped a tiny lace handkerchief.

Tesla was on his feet in an instant. He recovered the handkerchief, and with his hat in his other hand, bowing low from the waist, he handed the wisp of lace to the beautiful tragedienne, saying: "Madamoiselle, your handkerchief." Without even an upward glance at her graciously smiling face, he returned to his chair and resumed his conversation about his experiments on a world wireless system of power transmission.

When a newspaper reporter once asked Tesla why he had not married, his reply, as contained in the published interview was:

I have planned to devote my whole life to my work and for that reason I am denied the love and companionship of a good woman; and more, too.

I believe that a writer or a musician should marry. They gain inspiration that leads to finer achievement.

But an inventor has so intense a nature, with so much in it of wild, passionate quality that, in giving himself to a woman, he would give up everything, and so take everything from his chosen field. It is a pity, too; sometimes we feel so lonely.

In my student days I have known what it was to pass forty-eight hours at a stretch at a gaming table, undergoing intense emotion, that which most people believe is the strongest that can be known, but it is tame and insipid compared with that sublime moment when you see the labor of weeks fructify in a successful experiment that proves your theories. . . .

"Many times has Nikola Tesla known that supreme happiness," said the interviewer, "and he is likely to know it often again. It is impossible that his life work can be finished at forty. It would seem that his powers are only reaching their maturity."

Tesla was not unappreciative of the activities of the many women who showed a sincere interest in his welfare, and who tried to make life tolerable and pleasant for an obviously none-too-well-adjusted scientist projected into a social world from which he would have been only too willing to escape. He spoke glowingly of the first Mrs. Clarence Mackay (née Duer), Mrs. Jordan L. Mott, and of the beauty of Lady Ribblesdale (the former Mrs. John Jacob Astor). He admired the energetic idealism of Miss Anne Morgan; but never was the situation brightened by a single tint of romance.

He was impressed by the tall, graceful and charming Miss Marguerite Merington, a talented pianist and writer on musical subjects, who was a frequent dinner guest at the Johnson home.

"Why do you not wear diamonds and jewelry like other women?" Tesla undiplomatically asked Miss Merington, one evening.

"It is not a matter of choice with me," she replied, "but if I had enough money to load myself with diamonds I could think of better ways of spending it."

"What would you do with money if you had it?" the inventor continued.

"I would prefer to purchase a home in the country, except that I would not enjoy commuting to the suburbs," Miss Merington replied.

"Ah! Miss Merington, when I start getting my *millions* I will solve that problem. I will buy a square block here in New York and build a villa for you in the center and plant trees all around it. Then you will have your country home and will not have to leave the city."

Tesla was most generous in the distribution of his always still-to-be-gotten *millions;* none of his friends would ever have lacked anything they desired if he had had sufficient funds with which to satisfy their wishes. His promises, however, were always to be fulfilled—"When I start getting my *millions."*

Tesla had, as might be expected, very definite ideas about how women should dress. He also had clear-cut ideas about the feminine figure. He disliked the big "hefty" type and utterly detested fat women. The super-upholstered type, flashily dressed and heavily jeweled, that wasted time in hotel lobbies, were his pet abomination. He liked women trim, slim, graceful and agile.

One of his secretaries, well proportioned and a graceful blonde, wore to the office one day a dress that was in the very latest style. It was a summer dress made from a pretty print. The prevailing style called for an extremely low waist line, well down on the hips, several inches below its natural location. This gave a relatively short skirt and from the neck to the hips the dress was almost a plain cylinder. The style was very new, and was enjoying an intense but brief wave of popularity. The secretary was an excellent seamstress and had made the dress herself, an accomplishment of which she was justly proud.

Tesla summoned the secretary. She breezed into his sanctum not expecting, but hoping, that he would say something nice about her new dress.

"Miss," he said, "what is that you are wearing? You cannot wear that on this errand on which I wish you to go. I wished to have you take a note to a very important banker down town, and what would he think if someone from my office should come to him wearing such a monstrosity of a gown? How can you be such a slave to fashion? Whatever the fashion designers say is the style you buy and wear. Miss, you have good sense and good taste, so why did you let the saleslady in the store force a dress like this on you? Now if you were also very clever like my sister who makes all her own dresses you would not be forced to wear any such abominable style as this, then you too could make your own clothes and you could wear sensible gowns. You should always follow nature in the design of your clothes. Do not let a style designer deform nature for you, for then you become hideous instead of attractive. Now, Miss, you get into a cab, so not many people will see you, and go to your home and get into a sensible dress and return as soon as you can so you can take this letter down town for me."

Tesla never addressed any of his woman employees by either their Christian names or surnames. The only form of address he used to them was "Miss." As he spoke it, it sounded like "Meese," and he could make it very expressive. When he addressed the secretary wearing the gown of which he disapproved, it sounded like "Meeeeeeesssse." It could also be an abrupt, abbreviated expletive.

When a young woman on his office staff left his employ to get married, Tesla preached this sermonette to the remaining members:

"Do not marry too young. When you marry too young, men marry you mostly for your beauty and ten years later when your beauty is gone, they tire of you and become interested in someone else."

Tesla's attitude toward woman was paradoxical; he idealized woman—put her up on a pedestal—and yet he also viewed women in a purely objective and materialistic way, as if no spiritual concepts were involved in their make-up. This was undoubtedly an outward expression of the conflict that was taking place within his own life, between the normal healthy attitude toward female companionship, and the coldly objective planning of his life under which he refused to share the smallest fraction of his life with any woman.

Only the finest type of women could approach within friendship distance of Tesla, and such individuals were idealized by him without the least difficulty; he could desex them mentally

so that the vector or emotional attraction was eliminated. To the
remainder he did not bother to apply this process. They had no
attraction for him.

Out of the welter of human affairs, however, he visioned the
rising of a superior breed of human beings, few in number but
of vastly elevated intellectual status, while the remainder of the
race leveled itself on a merely productive and reproductive plane,
which, however, could represent a considerable improvement
over existing conditions. He sought to fashion an idealism out
of purely materialistic concepts of human nature. This was a
hold-over from the materialistic, agnostic views which were
fashionable and prevalent among scientists in the formative
period of his youth. This phase of his attitude was not particularly hard to break down in his latter years; but the phase which
represented an engineering approach to the solution of problems
of the human race was more firmly held, although he was willing to admit that spiritual factors had a real existence and should
be considered in such planning.

His views concerning women received their only expression
in published form in the article written for *Collier's,* in 1924,
by John B. Kennedy, from an interview with Tesla. On this
occasion, he said:

The struggle of the human female toward sex equality will end
up in a new sex order, with the females superior. The modern
woman, who anticipates in merely superficial phenomenon the advancement of her sex, is but a surface symptom of something deeper
and more potent fomenting in the bosom of the race.

It is not in the shallow physical imitation of the men that women
will assert first their equality and later their superiority, but in the
awakening of the intellect of women.

But the female mind has demonstrated a capacity for all the
mental acquirements and achievements of men, and as generations
ensue that capacity will be expanded; the average woman will be as
well educated as the average man, and then better educated, for the
dormant faculties of her brain will be stimulated into an activity
that will be all the more intense because of centuries of repose.

Women will ignore precedent and startle civilization with their progress.

The acquisition of new fields of endeavor by women, their gradual usurpation of leadership, will dull and finally dissipate feminine sensibilities, will choke the maternal instinct so that marriage and motherhood may become abhorrent and human civilization draw closer and closer to the perfect civilization of the bee.

The significance of this lies in the principle dominating the economy of the bee—the most highly organized and intelligently co-ordinated system of any form of non-rational animal life—the all governing supremacy of the instinct for immortality which makes divinity out of motherhood.

The center of all bee life is the queen. She dominates the hive, not through hereditary right, for any egg may be hatched into a reigning queen, but because she is the womb of the insect race.

There are vast desexualized armies of workers whose sole aim and business in life is hard work. It is the perfection of communism, of socialized, co-operative life wherein all things, including the young, are the common property of all.

Then there are the virgin bees, the princess bees, the females which are selected from the eggs of the queen when they are hatched and preserved in case an unfruitful queen should bring disappointment to the hive. And there are the male bees, few in number, unclean in habit, tolerated only because they are necessary to mate with the queen. . . .

The queen returns to the hive, impregnated, carrying with her tens of thousands of eggs—a future city of bees, and then begins the cycle of reproduction, the concentration of the teeming life of the hive in unceasing work for the birth of the new generation.

Imagination falters at the prospect of a human analogy to this mysterious and superbly dedicated civilization of the bee; but when we consider how the human instinct for race perpetuation dominates life in all its normal and exaggerated and perverse manifestations, there is ironic justice in the possibility that this instinct, with the intellectual advance of women, may be finally expressed after the manner of the bee, though it will take centuries to break down the habits and customs of peoples that bar the way to such a simply and scientifically ordered civilization.

If Tesla had been even half as well informed in the biological sciences as he was in the physical sciences, he probably would not have seen a possible solution of human problems in the social structure adapted to the limitations of an insect species which can never hope to utilize tools, and draw upon natural forces vastly exceeding their own energy sources, to work out their destiny. And more important is the fact that the bees can never hope to use advanced intellectual powers to improve their biological status, as can the human race. With a better knowledge of biological sciences he might have discovered that the physiological processes that control perpetuation of the individual are indissolubly linked to the processes that control the perpetuation of the race, and that by utilizing as much biological knowledge and spiritual insight, in designing a superman, as he utilized materialistic engineering principles, he might have designed himself as a more complete and potent superman, better adjusted to merging his intellectual creations into the current life of the race through a better understanding of human affairs.

Tesla tried to convince the world that he had succeeded in eliminating love and romance from his life; but he did not succeed. That failure (or perhaps from another aspect it was a success), is the story of the secret chapter of Tesla's life.

# TWENTY

THE most obvious outward characteristic of Tesla's life was his proclivity for feeding pigeons in public places. His friends knew he did it but never knew why. To the pedestrians on Fifth Avenue he was a familiar figure on the plazas of the Public Library at 42nd Street and St. Patrick's Cathedral at 50th Street. When he appeared and sounded a low whistle, the blue- and brown- and white-feathered flocks would appear from all directions, carpet the walks in front of him and even perch upon him while he scattered bird seed or permitted them to feed from his hand.

During the last three decades of his life, it is probable that not one out of tens of thousands who saw him knew who he was. His fame had died down and the generation that knew him well had passed on. Even when the newspapers, once a year, would break out in headlines about Tesla and his latest predictions concerning scientific wonders to come, no one associated that name with the excessively tall, very lean man, wearing clothes of a bygone era, who almost daily appeared to feed his feathered friends. He was just one of the strange individuals of whom it takes a great many of varying types to make up a complete population of a great metropolis.

When he started the practice, and no one knows just when that was, he was always dressed in the height of fashion and some of the world's most famous figures could frequently be seen in his company and joining him in scattering the bird seed, but there came a time when he paid less attention to his clothes, and those he wore became more and more old fashioned.

Fifth Avenue after midnight is a far different thoroughfare than the busy artery of human and vehicular traffic it is during the day. It is deserted. One can walk for blocks and meet no one except a policeman. On several occasions the author, by chance, met Tesla on an after-midnight walk up Fifth Avenue, going toward the library. Usually Tesla was quite willing to have one walk with him and chat upon a street encounter during the day, but on these after-midnight occasions he was definite about his desire to be left alone. "You will leave me now," he would say, bringing an abrupt end to a conversation hardly begun. The natural assumption was that Tesla was engaged on a definite line of thought and did not wish his mind to be diverted from its concentration on some knotty scientific problem. How far this was from the truth! And, as I learned much later, what a sacred significance these midnight pilgrimages to feed the pigeons—

which would come to his call, even from their nocturnal roosts—
had for him!

It was hard for almost everyone to understand why Tesla, engaged in momentous scientific developments, working twice as
many hours as the average individual, could see his way clear to
spend time scattering bird seed. The *Herald Tribune,* in an
editorial, once stated: "He would leave his experiments for a
time and feed the silly and inconsequential pigeons in Herald
Square."

It was a routine procedure in Tesla's office, however, for one
of his secretaries to go down town on a given day each week and
purchase three pounds each of rape, hemp and canary seed. This
was mixed in his office, and each day he took a small paper bag
filled with the seed and started on his rounds.

If, on any day, he was unable to make his pigeon-feeding
rounds, he would call a Western Union messenger boy, pay him
his fee, plus a dollar tip, and send him to feed the birds.

In addition to feeding the birds in the streets, Tesla took care
of pigeons in his rooms in the various hotels in which he made
his home. He usually had basket nests for from one to four
pigeons in his room and kept a cask of seed on hand to feed them.
The window to the room in which he kept these nests was never
closed.

Tesla became quite ill in his 40th Street office, one day in
1921. He was unable to work and lay upon his couch. As the
symptoms became more alarming and there was a possibility that
he might not be able to return to his room in the Hotel St. Regis,
he summoned his secretary to give her an "important" message.
As he spoke the important message, he required the secretary to
repeat each phrase after him to make sure that no errors would
be made. This required repetition was a usual procedure with
him; but in this case he was so ill, practically prostrate, that he
seemed hardly to have energy enough to speak the message a
single time.

"Miss," he whispered, "Call Hotel St. Regis—"

"Yes sir," she responded, "Call Hotel St. Regis—"

"Get the housekeeper on the fourteenth floor—"

"Get the housekeeper on the fourteenth floor—"

"Tell her to go to Mr. Tesla's room—"

"Tell her to go to Mr. Tesla's room—"

"And feed the pigeon today—"

"And feed the pigeon today—"

"The white female with touches of light gray in its wings—"

"The white female with touches of light gray in its wings—"

"And to continue doing this—"

"And to continue doing this—"

"Until she receives further orders from me—"

"Until she receives further orders from me—"

"There is plenty of feed in Mr. Tesla's room."

"There is plenty of feed in Mr. Tesla's room."

"Miss," he pleaded, "this is very important. Will you repeat the whole message to me so I can be sure you have it correct."

"Call Hotel St. Regis; get the housekeeper on the fourteenth floor. Tell her to go to Mr. Tesla's room and feed the pigeon today, the white female with touches of light gray on its wings, and continue doing this until she receives further orders from me. There is plenty of feed in Mr. Tesla's room."

"Ah, yes," said Tesla, his eyes brightening as he spoke, "the white one with touches of light gray in its wings. And if I am not here tomorrow, you will repeat that message then and every day until you get my further orders. Do it now, Miss—it is very important."

Tesla's orders were always carried out to the letter and this one particularly, since he had placed such unusual emphasis on it. His secretary and the members of his staff felt that his illness must be more serious than it seemed to be, since at a time when he had a great many very serious problems on his hands and he appeared to be on the verge of a siege of illness, the more pressing situations were completely forgotten and his only thought was of a pigeon. He must be delirious, so they thought.

Some months later Tesla failed one day to show up at his office, and when his secretary telephoned to his hotel, the inventor informed her that he was all right, but that his pigeon was ill and he dared not leave the room for fear she would need him. He remained in his room for several days.

About a year later Tesla came to his office earlier than usual one day, and apparently very much disturbed. He carried a small bundle in a tender manner on his bent arm. He telephoned to Julius Czito, a machinist on whom he frequently depended to perform unusual tasks, and asked him to come to the office. Czito lived in the suburbs. He told him briefly that the bundle contained a pigeon that had died in his room at the hotel, and that he desired to have it properly buried on Czito's property where the grave could be cared for. Czito, in relating the incident years afterward, said he was tempted, on leaving the office, to drop the package in the first garbage can he found; but something caused him to desist and he took it to his home. Before he could perform the burial, Tesla telephoned to his home and asked him to return the package the next morning. How Tesla disposed of it is not known.

In 1924 Tesla's financial condition fell to a very low level. He was completely broke. He was unable to pay his rent and there were some judgments against him for other unpaid bills. A deputy sheriff appeared at his office one afternoon to seize everything in the office to satisfy a judgment. Tesla managed to talk the sheriff into delaying seizure. When the official had gone he took stock of his situation. He had not paid his secretaries' wages for two weeks and he now owed them for another fraction of a week. He was entirely without funds in the bank. A search of his safe disclosed that the only object of negotiable value was the heavy gold Edison Medal presented to him by the American Institute of Electrical Engineers in 1917.

"Miss, and Miss," he said, addressing the secretaries. "This medal contains about one hundred dollars' worth of gold. I will have it cut in half and give each of you one-half, or one of you can take all of it and I will later pay the other."

The two young women, Miss Dorothy F. Skerritt and Miss Muriel Arbus, refused to permit him either to damage or part with the medal, and offered instead to aid him with the meager amounts of cash they had in their purses, which offer he refused with thanks. (A few weeks later the girls received their back salaries, at $35 per week, and an additional two weeks' salary.)

A search of the cash drawer revealed a little over $5.00—all the money he possessed.

"Ah! Miss," he said, "that will be enough to buy the bird seed. I am all out of seed, so will you go down town in the morning and purchase some and deliver it to my hotel."

Again calling his trusted aide, Czito (whom he was forced to leave unpaid to the extent of $1,100), he put up to him the problem of vacating the office immediately. Within a few hours the entire contents of the offices were stored in a near-by office building.

A short time later he was forced to leave his apartment in the Hotel St. Regis. His bill had been unpaid for some time, but the immediate cause was associated with pigeons. He had been spending more time in his hotel room, which also became his office, and devoted more time to feeding pigeons. Great flocks of them would come to his windows and into the rooms, and their dirt on the outside of the building became a problem to the management and on the inside to the maids. He sought to solve the problem by putting the birds in a hamper and having George Scherff take them to his Westchester home. Three weeks later, when first given their freedom, they returned, one making the trip in half an hour. Tesla was given his choice of ceasing to feed the pigeons or leaving the hotel. He left.

He next made his home at the Hotel Pennsylvania. He remained there a few years and the same situation, both as to bills and pigeons, developed. He moved to the Hotel Governor Clinton—and in about a year went through the same experience.

He next moved to the Hotel New Yorker, in 1933, where he spent the final ten years of his life.

After midnight one night in the fall of 1937, Tesla started out from the Hotel New Yorker to make his regular pilgrimage to the Cathedral and the Library to feed the pigeons. In crossing a street a couple of blocks from the hotel an accident happened, how is unknown. In spite of his agility, he was unable to avoid contact with a moving taxicab, and was thrown heavily to the ground. He raised no question as to who was at fault, refused medical aid, and asked merely to be taken to his hotel in another cab.

Arriving at the hotel, he went to bed and had scarcely got under the covers when he telephoned for his favorite messenger boy, Kerrigan, from a near-by Western Union office, gave him the package of bird seed and directed him to complete the task which he had started and the accident interrupted.

The next day, when it was apparent that he would be unable to take his usual daily walks for some time to come, he hired the messenger for six months to feed the pigeons every day. Tesla's back had been severely wrenched in the accident, and three ribs broken, but the full extent of his injuries will never be known for, in keeping with his almost lifelong custom, he refused to consult a doctor. Pneumonia developed but for this he also refused medical aid. He was bedridden for some months, and was unable to carry on his practice of feeding pigeons from his window; and soon they failed to come.

In the spring of 1938 he was able to get up. He at once resumed his pigeon-feeding walks on a much more limited scale, but frequently had a messenger act for him.

This devotion to his pigeon-feeding task seemed to everyone who knew him like nothing more than the hobby of an eccentric scientist, but if they could have looked into Tesla's heart, or read his mind, they would have discovered that they were witnessing the world's most fantastic, yet tender and pathetic love affair.

TESLA, as a self-made superman, suffered from the limitations of his maker. Endowed with an intelligence above the average in both quality and quantity, and with some supernormal faculties, he was able to erect a superman higher in stature than himself; but the greater height was attained by sacrificing other dimensions, and in this diminution of breadth and thickness existed a deficiency.

When he was a youth and his mind was in its most plastic and formative stage, he adopted, as we have seen, the then prevalent agnostic and materialistic view of life. Today science has emancipated itself from slavery to either an antagonistic mysticism or materialism, and is willing to consider both as harmonious parts of a comprehensive approach to the understanding of Nature, but is conscious that it has not yet learned how to manipulate or control the more intangible factors upon which the mystics have builded their structures of knowledge. Vast realms of human experience have been rejected in all ages by scientists, of whatever name, who failed to fit them in logical arrangement in their inadequate and too simplified natural philosophies. By rejecting the phenomena that lay beyond their intellectual abilities, the scientists and philosophers did not eliminate them nor prevent their manifestations. The phenomena so rejected, however, were given an academic home by the ecclesiasts, who accepted them without understanding, or hope of understanding, and thus incarcerated them in the foundation of the religious mysteries where they served a useful purpose, for upon an unknown it is possible to build a greater unknown.

The mystical experiences of the saints, of whatever faith, are demonstrations of forces which are natural functions of the phenomenon of life, expressed in varying degree in step with the expanding unfoldment of the individual toward an advanced state of evolution.

Tesla was an individual in an advanced state of development, and there came to him experiences which he refused to accept as experiments; accepting the benefits which came to him but rejecting the vehicle which transported them. This was true, for example, in the case of the burst of revelation which came to him —revealing scores of tremendously valuable inventions—while he strolled in the park at Budapest, and which differed only in degree and type, but not in fundamental nature, from the blinding light which came to Saul on the road to Damascus, and to others to whom *illumination* has come by similar processes.

His materialistic concepts made him intellectually blind to the strange phenomenon by which revelation, or illumination, had come to him, but made him more keenly appreciative of the value of that which was revealed. It must not be understood that this revelation was a happenstance phenomenon of the moment, for Tesla, endowed by Nature with an intellect capable of vast unfoldment, had exerted almost superhuman efforts to achieve that which was revealed to him, and the effort was not unassociated with the result.

In a contrary direction, Tesla suppressed a tremendously large or important realm of his life by the planned elimination of love and romance from his thoughts and experience. Just as his efforts to discover the physical secrets of Nature built up forces that penetrated to the plane of revelation, so did his equally tremendous effort to suppress love and romance build up forces, beyond his control, that were operating to express themselves. There was a parallel situation in his philosophy of natural phenomena, in that he suppressed all spiritual aspects of Nature and confined himself to the purely materialistic aspects.

Two forces, one of love and romance in his personal nature, and the other the spiritual aspects of Nature in his philosophy, as applied to his work, were incarcerated in a limbo of his personality, seeking an outlet into the paradise of expression and manifestation. And they obtained that outlet, expressing their nature by the form of the manifestation; but Tesla failed to recognize them. Tesla, rejecting the love of woman and thinking that he had engineered a complete elimination of the problem of love, failed to excise from his nature the capacity to love, and when this capacity expressed itself, it did so by directing its energies through a channel he left unguarded in planning the self-made superman.

The manifestation of these united forces of love and spirituality resulted in a fantastic situation, probably without parallel in human annals. Tesla told me the story; but if I did not have a witness who assured me that he heard exactly what I heard, I would have convinced myself that I had had nothing more tangible than a dream experience. It was the love story of Tesla's life. In the story of his strange romance, I saw instantly the reason for those unremitting daily journeys to feed the pigeons, and those midnight pilgrimages when he wished to be alone. I recalled those occasions when I had happened to meet him on deserted Fifth Avenue and, when I spoke to him, he replied, "You will now leave me." He told his story simply, briefly and without embellishments, but there was still a surging of emotion in his voice.

"I have been feeding pigeons, thousands of them, for years; thousands of them, for who can tell—

"But there was one pigeon, a beautiful bird, pure white with light gray tips on its wings; that one was different. It was a female. I would know that pigeon anywhere.

"No matter where I was that pigeon would find me; when I wanted her I had only to wish and call her and she would come flying to me. She understood me and I understood her.

"I loved that pigeon.

"Yes," he replied to an unasked question. "Yes, I loved that pigeon, I loved her as a man loves a woman, and she loved me. When she was ill I knew, and understood; she came to my room and I stayed beside her for days. I nursed her back to health. That pigeon was the joy of my life. If she needed me, nothing else mattered. As long as I had her, there was a purpose in my life.

"Then one night as I was lying in my bed in the dark, solving problems, as usual, she flew in through the open window and

*316*

stood on my desk. I knew she wanted me; she wanted to tell me something important so I got up and went to her.

"As I looked at her I knew she wanted to tell me—she was dying. And then, as I got her message, there came a light from her eyes—powerful beams of light.

"Yes," he continued, again answering an unasked question, "it was a real light, a powerful, dazzling, blinding light, a light more intense than I had ever produced by the most powerful lamps in my laboratory.

"When that pigeon died, something went out of my life. Up to that time I knew with a certainty that I would complete my work, no matter how ambitious my program, but when that something went out of my life I knew my life's work was finished.

"Yes, I have fed pigeons for years; I continue to feed them, thousands of them, for after all, who can tell—"

There was nothing more to say. We parted in silence. The talk took place in a corner of the mezzanine in the Hotel New Yorker. I was accompanied by William L. Laurence, science writer of the New York *Times*. We walked several blocks on Seventh Avenue before we spoke.

No longer was there any mystery to the midnight pilgrimages when he called the pigeons from their niches in the Gothic tracery of the Cathedral, or from under the eaves of the Greek temple that houses the Library—pursuing, among the thousands of them . . . "For after all, who can tell . . . ?"

It is out of phenomena such as Tesla experienced when the dove flew out of the midnight darkness and into the blackness of his room and flooded it with blinding light, and the revelation that came to him out of the dazzling sun in the park at Budapest, that the mysteries of religion are built. But he comprehended them not; for, if he had not suppressed the rich mystical inheritance of his ancestors that would have brought enlightenment, he would have understood the symbolism of the Dove.

# JOHN J. O'NEILL

Editor; b. New York, N.Y., June 21, 1889; s. James and Catherine (Kelleher) O'N.; ed. public schools and night school, New York, N.Y., and Internat. Corr. Schools; m. Marie Bock, July 7, 1912; children — Kenneth Horace, Peggy Theresa (Mrs. Clyde T. Grayson). Began as printer, 1903-04; electrician, 1905-06; with N.Y. (Astor) Public Library, 1907-08; reporter, 1908-15; reporter, Brooklyn (N.Y.) Daily Eagle, 1915-17, feature editor, 1918-22, radio editor, 1922-25, automobile and aviation editor, 1925-26, science editor, 1926-32, supervisor construction of new building and plant, 1929-30; science editor New York Herald Tribune since 1933. Organizer Suffolk County Home Defense Regt., 1917. Newspapermen's Officers Training Corps, 1917. Served as pvt., Machine Gun Co., 7th Regt., N.Y. Nat. Guard, 1917-19. Recipient of Pulitzer Award in Journalism, Columbia U., 1937. Best Science Story of Year, U. of Kan., 1938; Clement Cleveland Award in Journalism, Columbia U., 1937; Best Science Story of Year, U. of Kan., 1938; Clement Cleveland Award (shared), N.Y. Cancer Soc., 1938; Westinghouse Distinguished Science Writing Medal from A.A.A.S., 1946. Revealed that atomic energy had been released, Mar. 1940. Mem. Arts and Science Conf., 1937. Fellow Am. Geog. Soc., Arctic Institute of North America; mem. Amateur Astronomers Association, A.A.A.S., Am. Genetic Assn., Am. Inst. City of N.Y. (chmn. bd. mgrs., 1933-37), Am. Soc. Psychical Research (chmn. research com. and trustee, 1933-37), Am. Polar Soc., India-Am. Science Assn. (founding mem.), Am. Acad. Polit. and Social Science, Acad. Polit. Sci., Am.-Irish Hist. Soc., Am. Newspaper Guild (founding mem.), Am. Rocket Soc., Am. Oriental Soc., Am. Meteorol. Soc., Am. Geophys. Union, Nat. Assn. Science Writers (charter mem.; v.p., 1939, pres., 1940); Royal Astron. Soc. (Canada), Assn. Lunar and Obs., Astron. Society of Pacific, Am. Assn. Physics Teachers. Authro: Enter Atomic Energy, 1940; Prodigal Genius, The Life of Nikola Tesla, 1944; Almighty Atom, The Real Story of Atomic Energy, 1945; You and the Universe, 1945; Engineering The New Age, 1949. Contbr. tech. articles to sci. publs. Home: 200 N. Long Beach Av., Freeport, L.I., N.Y. Office 230 W. 41st St., New York, N.Y. Died Aug. 30, 1953.

# ACKNOWLEDGMENTS

MUCH valuable aid has been received from many sources in the preparation of this volume. For this helpful co-operation my thanks are due to:

Sava N. Kosanovic, Minister of State of Yugoslavia, and Tesla's nephew, for making available books, family records, transcripts of records, pictures, and for correcting the manuscript of many chapters; and to his secretary, Miss Charlotte Muzar;

Miss Dorothy Skerritt and Miss Muriel Arbus, Tesla's secretaries; and George Scherff and Julius C. Czito, business associates;

Mrs. Margaret C. Behrend, for the privilege of reading correspondence between her husband and Tesla; and to Dr. W. B. Earle, Dean of Engineering, Clemson Agricultural College, for pictures and other material from the Behrend Collection in the college library;

Mrs. Agnes Holden, daughter of the late Robert Underwood Johnson, ambassador, and editor of the *Century Magazine;* Miss Marguerite Merington; Mrs. Grizelda M. Hobson, widow of the late Rear Admiral Hobson; Waldemar Kaempffert, Science Editor of the New York *Times;* Professor Emeritus Charles F. Scott, Department of Electrical Engineering, Yale University; Hans Dahlstrand, of the Allis Chalmers Manufacturing Co.; Leo Maloney, Manager of the Hotel New Yorker; and W. D. Crow, architect of the Tesla tower, for reminiscences, data, or helpful conversations concerning their contacts with Tesla;

## ACKNOWLEDGMENTS

Florence S. Hellman, Chief of the Bibliographic Division, Library of Congress; Olive E. Kennedy, Research Librarian of the Public Information Center, National Electric Manufacturers Association; A. P. Peck, Managing Editor of the *Scientific American;* Myrta L. Mason, and Charles F. Pflaging, for bibliographic aid;

G. Edward Pendray and his associates in the Westinghouse Electric and Manufacturing Co., and C. D. Wagoner and his associates in the General Electric Co., for correcting, or reading and making helpful suggestions in connection with many chapters;

William L. Laurence, science writer of the New York *Times,* and Bloyce Fitzgerald, for exchange of data;

Randall Warden; William Spencer Bowen, President of the Bowen Research Corp.; G. H. Clark, of the Radio Corporation of America; Kenneth M. Swezey, of *Popular Science;* Mrs. Mabel Fleischer and Carl Payne Tobey, who have aided in a variety of ways;

*Collier's*—The National Weekly; *The American Magazine;* the New York *World-Telegram* and the General Electric Co., for permission to quote copyrighted material, for which credit is given where quoted; and

Peggy O'Neill Grayson, my daughter, for extended secretarial service.

To all the foregoing I extend my sincere thanks.

JOHN J. O'NEILL

*Freeport, L. I.*
*New York*
*July 15, 1944*

# PARTIAL LIST OF TESLA U. S. PATENTS

---

Direct Current. Motor and generator controls, Arc lights, etc.:

334,823; 335,786; 335,787; 336,961; 336,962; 350,954; 382,845.

Polyphase Currents. Electric transmissions of power, Dynamos, Motors, Transformers, Electrical Distribution:

395,748; 381,968; 381,969; 381,970; 382,279; 382,280;
382,281; 382,282; 390,413; 390,414; 390,415; 390,721;
390,820; 401,520; 405,858; 405,859; 406,968; 413,353;
413,702; 413,703; 416,191; 416,192; 416,193; 416,194;
416,195; 417,794; 418,248; 424,036; 433,700; 433,701;
445,207; 455,067; 459,772; 464,666; 487,796; 511,559;
511,560; 511,915; 512,340; 524,426; 555,190.

Currents of High Frequency and High Potential. Apparatus for generating, controls, circuits and systems, insulation, and applications:

314,167; 314,168; 447,920; 447,921; 454,622; 455,069;
462,418; 464,667; 511,916; 514,170; 567,818; 568,176;
568,177; 568,178; 568,179; 568,180; 568,671; 577,670;
583,953; 609,245; 609,246; 609,247; 609,248; 609,249;
609,250; 609,251; 611,719; 613,735.

Wireless Systems. Radio telegraphy, Radio mechanics, Methods of tuning and selection, Detectors, etc.:

568,178;   593,138;   613,809;   645,576;   649,621;   655,838;
685,012;   685,953;   685,954;   685,955;   685,956;   685,957;
685,958;   723,188;   725,605;   787,412;   1,119,732.

Various patents. Steam turbines, Pumps, Speedometers, Airplanes, Mechanical oscillators, Thermo-magnetic motors, etc.:

396,121;      428,057;      455,068;      514,169;      517,900;
524,972;      524,973;      1,061,142;    1,113,716;    1,209,359;
1,266,175;    1,274,816;    1,314,718;    1,329,559;    1,365,547;
1,402,025;    1,655,114.

# INDEX

# INDEX

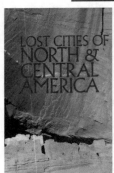
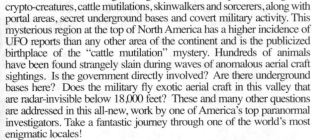
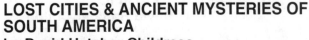

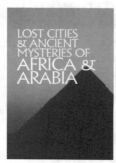

## LOST CITIES & ANCIENT MYSTERIES OF AFRICA & ARABIA
### by David Hatcher Childress

Childress continues his world-wide quest for lost cities and ancient mysteries. Join him as he discovers forbidden cities in the Empty Quarter of Arabia; "Atlantean" ruins in Egypt and the Kalahari desert; a mysterious, ancient empire in the Sahara; and more. This is the tale of an extraordinary life on the road: across war-torn countries, Childress searches for King Solomon's Mines, living dinosaurs, the Ark of the Covenant and the solutions to some of the fantastic mysteries of the past.

**423 PAGES. 6x9 PAPERBACK. ILLUSTRATED. $14.95. CODE: AFA**

## LOST CITIES OF ATLANTIS, ANCIENT EUROPE & THE MEDITERRANEAN
### by David Hatcher Childress

Childress takes the reader in search of sunken cities in the Mediterranean; across the Atlas Mountains in search of Atlantean ruins; to remote islands in search of megalithic ruins; to meet living legends and secret societies. From Ireland to Turkey, Morocco to Eastern Europe, and around the remote islands of the Mediterranean and Atlantic, Childress takes the reader on an astonishing quest for mankind's past. Ancient technology, cataclysms, megalithic construction, lost civilizations and devastating wars of the past are all explored in this book.

**524 PAGES. 6x9 PAPERBACK. ILLUSTRATED. $16.95. CODE: MED**

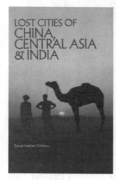

## LOST CITIES OF CHINA, CENTRAL ASIA & INDIA
### by David Hatcher Childress

Like a real life "Indiana Jones," maverick archaeologist David Childress takes the reader on an incredible adventure across some of the world's oldest and most remote countries in search of lost cities and ancient mysteries. Discover ancient cities in the Gobi Desert; hear fantastic tales of lost continents, vanished civilizations and secret societies bent on ruling the world; visit forgotten monasteries in forbidding snow-capped mountains with strange tunnels to mysterious subterranean cities! A unique combination of far-out exploration and practical travel advice, it will astound and delight the experienced traveler or the armchair voyager.

**429 PAGES. 6x9 PAPERBACK. ILLUSTRATED. FOOTNOTES & BIBLIOGRAPHY. $14.95. CODE: CHI**

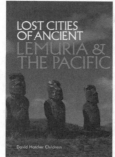

## LOST CITIES OF ANCIENT LEMURIA & THE PACIFIC
### by David Hatcher Childress

Was there once a continent in the Pacific? Called Lemuria or Pacifica by geologists, Mu or Pan by the mystics, there is now ample mythological, geological and archaeological evidence to "prove" that an advanced and ancient civilization once lived in the central Pacific. Maverick archaeologist and explorer David Hatcher Childress combs the Indian Ocean, Australia and the Pacific in search of the surprising truth about mankind's past. Contains photos of the underwater city on Pohnpei; explanations on how the statues were levitated around Easter Island in a clockwise vortex movement; tales of disappearing islands; Egyptians in Australia; and more.

**379 PAGES. 6x9 PAPERBACK. ILLUSTRATED. FOOTNOTES & BIBLIOGRAPHY. $14.95. CODE: LEM**

# A HITCHHIKER'S GUIDE TO ARMAGEDDON
## by David Hatcher Childress

With wit and humor, popular Lost Cities author David Hatcher Childress takes us around the world and back in his trippy finalé to the Lost Cities series. He's off on an adventure in search of the apocalypse and end times. Childress hits the road from the fortress of Megiddo, the legendary citadel in northern Israel where Armageddon is prophesied to start. Hitchhiking around the world, Childress takes us from one adventure to another, to ancient cities in the deserts and the legends of worlds before our own. In the meantime, he becomes a cargo cult god on a remote island off New Guinea, gets dragged into the Kennedy Assassination by one of the "conspirators," investigates a strange power operating out of the Altai Mountains of Mongolia, and discovers how the Knights Templar and their off-shoots have driven the world toward an epic battle centered around Jerusalem and the Middle East.

320 PAGES. 6x9 PAPERBACK. ILLUSTRATED. BIBLIOGRAPHY. INDEX. $16.95. CODE: HGA

# TECHNOLOGY OF THE GODS
## The Incredible Sciences of the Ancients
## by David Hatcher Childress

Childress looks at the technology that was allegedly used in Atlantis and the theory that the Great Pyramid of Egypt was originally a gigantic power station. He examines tales of ancient flight and the technology that it involved; how the ancients used electricity; megalithic building techniques; the use of crystal lenses and the fire from the gods; evidence of various high tech weapons in the past, including atomic weapons; ancient metallurgy and heavy machinery; the role of modern inventors such as Nikola Tesla in bringing ancient technology back into modern use; impossible artifacts; and more.

356 PAGES. 6x9 PAPERBACK. ILLUSTRATED. BIBLIOGRAPHY. $16.95. CODE: TGOD

# VIMANA AIRCRAFT OF ANCIENT INDIA & ATLANTIS
## by David Hatcher Childress, introduction by Ivan T. Sanderson

In this incredible volume on ancient India, authentic Indian texts such as the *Ramayana* and the *Mahabharata* are used to prove that ancient aircraft were in use more than four thousand years ago. Included in this book is the entire Fourth Century BC manuscript *Vimaanika Shastra* by the ancient author Maharishi Bharadwaaja. Also included are chapters on Atlantean technology, the incredible Rama Empire of India and the devastating wars that destroyed it.

334 PAGES. 6x9 PAPERBACK. ILLUSTRATED. $15.95. CODE: VAA

# LOST CONTINENTS & THE HOLLOW EARTH
## I Remember Lemuria and the Shaver Mystery
## by David Hatcher Childress & Richard Shaver

Shaver's rare 1948 book *I Remember Lemuria* is reprinted in its entirety, and the book is packed with illustrations from Ray Palmer's *Amazing Stories* magazine of the 1940s. Palmer and Shaver told of tunnels running through the earth—tunnels inhabited by the Deros and Teros, humanoids from an ancient spacefaring race that had inhabited the earth, eventually going underground, hundreds of thousands of years ago. Childress discusses the famous hollow earth books and delves deep into whatever reality may be behind the stories of tunnels in the earth. Operation High Jump to Antarctica in 1947 and Admiral Byrd's bizarre statements, tunnel systems in South America and Tibet, the underground world of Agartha, the belief of UFOs coming from the South Pole, more.

344 PAGES. 6x9 PAPERBACK. ILLUSTRATED. $16.95. CODE: LCHE

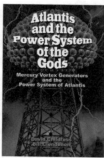

## ATLANTIS & THE POWER SYSTEM OF THE GODS
### Mercury Vortex Generators & the Power System of Atlantis
### by David Hatcher Childress and Bill Clendenon

Childress' fascinating analysis of Nikola Tesla's broadcast system in light of Edgar Cayce's "Terrible Crystal" and the obelisks of ancient Egypt and Ethiopia. Includes: Atlantis and its crystal power towers that broadcast energy; how these incredible power stations may still exist today; inventor Nikola Tesla's nearly identical system of power transmission; Mercury Proton Gyros and mercury vortex propulsion; more. Richly illustrated, and packed with evidence that Atlantis not only existed—it had a world-wide energy system more sophisticated than ours today.

**246 PAGES. 6x9 PAPERBACK. ILLUSTRATED. $15.95. CODE: APSG**

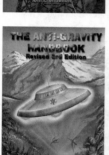

## THE ANTI-GRAVITY HANDBOOK
### edited by David Hatcher Childress, with Nikola Tesla, T.B. Paulicki, Bruce Cathie, Albert Einstein and others

The new expanded compilation of material on Anti-Gravity, Free Energy, Flying Saucer Propulsion, UFOs, Suppressed Technology, NASA Cover-ups and more. Highly illustrated with patents, technical illustrations and photos. This revised and expanded edition has more material, including photos of Area 51, Nevada, the government's secret testing facility. This classic on weird science is back in a new format!

**230 PAGES. 7x10 PAPERBACK. ILLUSTRATED. $16.95. CODE: AGH**

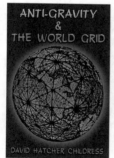

## ANTI–GRAVITY & THE WORLD GRID

Is the earth surrounded by an intricate electromagnetic grid network offering free energy? This compilation of material on ley lines and world power points contains chapters on the geography, mathematics, and light harmonics of the earth grid. Learn the purpose of ley lines and ancient megalithic structures located on the grid. Discover how the grid made the Philadelphia Experiment possible. Explore the Coral Castle and many other mysteries, including acoustic levitation, Tesla Shields and scalar wave weaponry. Browse through the section on anti-gravity patents, and research resources.

**274 PAGES. 7x10 PAPERBACK. ILLUSTRATED. $14.95. CODE: AGW**

## ANTI–GRAVITY & THE UNIFIED FIELD
### edited by David Hatcher Childress

Is Einstein's Unified Field Theory the answer to all of our energy problems? Explored in this compilation of material is how gravity, electricity and magnetism manifest from a unified field around us. Why artificial gravity is possible; secrets of UFO propulsion; free energy; Nikola Tesla and anti-gravity airships of the 20s and 30s; flying saucers as superconducting whirls of plasma; anti-mass generators; vortex propulsion; suppressed technology; government cover-ups; gravitational pulse drive; spacecraft & more.

**240 PAGES. 7x10 PAPERBACK. ILLUSTRATED. $14.95. CODE: AGU**

## MAPS OF THE ANCIENT SEA KINGS
### Evidence of Advanced Civilization in the Ice Age
### by Charles H. Hapgood

Charles Hapgood has found the evidence in the Piri Reis Map that shows Antarctica, the Hadji Ahmed map, the Oronteus Finaeus and other amazing maps. Hapgood concluded that these maps were made from more ancient maps from the various ancient archives around the world, now lost. Not only were these unknown people more advanced in mapmaking than any people prior to the 18th century, it appears they mapped all the continents. The Americas were mapped thousands of years before Columbus. Antarctica was mapped when its coasts were free of ice!

316 PAGES. 7x10 PAPERBACK. ILLUSTRATED. BIBLIOGRAPHY & INDEX. $19.95. CODE: MASK

## PATH OF THE POLE
### Cataclysmic Pole Shift Geology
### by Charles H. Hapgood

*Maps of the Ancient Sea Kings* author Hapgood's classic book *Path of the Pole* is back in print! Hapgood researched Antarctica, ancient maps and the geological record to conclude that the Earth's crust has slipped on the inner core many times in the past, changing the position of the pole. *Path of the Pole* discusses the various "pole shifts" in Earth's past, giving evidence for each one, and moves on to possible future pole shifts.

356 PAGES. 6x9 PAPERBACK. ILLUSTRATED. $16.95. CODE: POP

## SECRETS OF THE HOLY LANCE
### The Spear of Destiny in History & Legend
### by Jerry E. Smith

*Secrets of the Holy Lance* traces the Spear from its possession by Constantine, Rome's first Christian Caesar, to Charlemagne's claim that with it he ruled the Holy Roman Empire by Divine Right, and on through two thousand years of kings and emperors, until it came within Hitler's grasp—and beyond! Did it rest for a while in Antarctic ice? Is it now hidden in Europe, awaiting the next person to claim its awesome power? Neither debunking nor worshiping, *Secrets of the Holy Lance* seeks to pierce the veil of myth and mystery around the Spear. Mere belief that it was infused with magic by virtue of its shedding the Savior's blood has made men kings. But what if it's more? What are "the powers it serves"?

312 PAGES. 6x9 PAPERBACK. ILLUSTRATED. BIBLIOGRAPHY. $16.95. CODE: SOHL

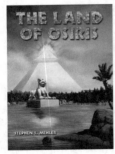

## THE LAND OF OSIRIS
### An Introduction to Khemitology
### by Stephen S. Mehler

Was there an advanced prehistoric civilization in ancient Egypt who built the great pyramids and carved the Great Sphinx? Did the pyramids serve as energy devices and not as tombs for kings? Mehler has uncovered an indigenous oral tradition that still exists in Egypt, and has been fortunate to have studied with a living master of this tradition, Abd'El Hakim Awyan. Mehler has also been given permission to present these teachings to the Western world, teachings that unfold a whole new understanding of ancient Egypt . Chapters include: Egyptology and Its Paradigms; Asgat Nefer—The Harmony of Water; Khemit and the Myth of Atlantis; The Extraterrestrial Question; more.

272 PAGES. 6x9 PAPERBACK. ILLUSTRATED. COLOR SECTION. BIBLIOGRAPHY. $18.00 CODE: LOOS

# REICH OF THE BLACK SUN
## Nazi Secret Weapons & the Cold War Allied Legend
### by Joseph P. Farrell

Why were the Allies worried about an atom bomb attack by the Germans in 1944? Why did the Soviets threaten to use poison gas against the Germans? Why did Hitler in 1945 insist that holding Prague could win the war for the Third Reich? Why did US General George Patton's Third Army race for the Skoda works at Pilsen in Czechoslovakia instead of Berlin? Why did the US Army not test the uranium atom bomb it dropped on Hiroshima? Why did the Luftwaffe fly a non-stop round trip mission to within twenty miles of New York City in 1944? *Reich of the Black Sun* takes the reader on a scientific-historical journey in order to answer these questions. Arguing that Nazi Germany actually won the race for the atom bomb in late 1944,

352 PAGES. 6x9 PAPERBACK. ILLUSTRATED. BIBLIOGRAPHY.
**$16.95. CODE: ROBS**

# THE GIZA DEATH STAR
## The Paleophysics of the Great Pyramid & the Military Complex at Giza
### by Joseph P. Farrell

Was the Giza complex part of a military installation over 10,000 years ago? Chapters include: An Archaeology of Mass Destruction, Thoth and Theories; The Machine Hypothesis; Pythagoras, Plato, Planck, and the Pyramid; The Weapon Hypothesis; Encoded Harmonics of the Planck Units in the Great Pyramid; High Frequency Direct Current "Impulse" Technology; The Grand Gallery and its Crystals: Gravito-acoustic Resonators; The Other Two Large Pyramids; the "Causeways," and the "Temples"; A Phase Conjugate Howitzer; Evidence of the Use of Weapons of Mass Destruction in Ancient Times; more.

290 PAGES. 6x9 PAPERBACK. ILLUSTRATED. $16.95. CODE: GDS

# THE GIZA DEATH STAR DEPLOYED
## The Physics & Engineering of the Great Pyramid
### by Joseph P. Farrell

Farrell expands on his thesis that the Great Pyramid was a maser, designed as a weapon and eventually deployed—with disastrous results to the solar system. Includes: Exploding Planets: A Brief History of the Exoteric and Esoteric Investigations of the Great Pyramid; No Machines, Please!; The Stargate Conspiracy; The Scalar Weapons; Message or Machine?; A Tesla Analysis of the Putative Physics and Engineering of the Giza Death Star; Cohering the Zero Point, Vacuum Energy, Flux: Feedback Loops and Tetrahedral Physics; and more.

290 PAGES. 6x9 PAPERBACK. ILLUSTRATED. $16.95. CODE: GDSD

# THE GIZA DEATH STAR DESTROYED
## The Ancient War For Future Science
### by Joseph P. Farrell

Recapping his earlier books, Farrell moves on to events of the final days of the Giza Death Star and its awesome power. These final events, eventually leading up to the destruction of this giant machine, are dissected one by one, leading us to the eventual abandonment of the Giza Military Complex—an event that hurled civilization back into the Stone Age. Chapters include: The Mars-Earth Connection; The Lost "Root Races" and the Moral Reasons for the Flood; The Destruction of Krypton: The Electrodynamic Solar System, Exploding Planets and Ancient Wars; Turning the Stream of the Flood: the Origin of Secret Societies and Esoteric Traditions; The Quest to Recover Ancient Mega-Technology; Vortices and Mass Particles: The Topology of the Aether; "Acoustic" Intensity of Fields; The Pyramid of Crystals; tons more.

**292 pages. 6x9 paperback. Illustrated. $16.95. Code: GDES**

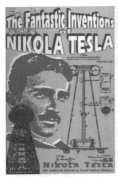

## THE FANTASTIC INVENTIONS OF NIKOLA TESLA
### by Nikola Tesla with additional material by David Hatcher Childress

This book is a readable compendium of patents, diagrams, photos and explanations of the many incredible inventions of the originator of the modern era of electrification. In Tesla's own words are such topics as wireless transmission of power, death rays, and radio-controlled airships. In addition, rare material on a secret city built at a remote jungle site in South America by one of Tesla's students, Guglielmo Marconi. Marconi's secret group claims to have built flying saucers in the 1940s and to have gone to Mars in the early 1950s! Incredible photos of these Tesla craft are included. •His plan to transmit free electricity into the atmosphere. •How electrical devices would work using only small antennas. •Why unlimited power could be utilized anywhere on earth. •How radio and radar technology can be used as death-ray weapons in Star Wars.

**342 PAGES. 6x9 PAPERBACK. ILLUSTRATED. $16.95. CODE: FINT**

## THE TESLA PAPERS
### Nikola Tesla on Free Energy & Wireless Transmission of Power
### by Nikola Tesla, edited by David Hatcher Childress

David Hatcher Childress takes us into the incredible world of Nikola Tesla and his amazing inventions. Tesla's fantastic vision of the future, including wireless power, anti-gravity, free energy and highly advanced solar power. Also included are some of the papers, patents and material collected on Tesla at the Colorado Springs Tesla Symposiums, including papers on: •The Secret History of Wireless Transmission •Tesla and the Magnifying Transmitter •Design and Construction of a Half-Wave Tesla Coil •Electrostatics: A Key to Free Energy •Progress in Zero-Point Energy Research •Electromagnetic Energy from Antennas to Atoms •Tesla's Particle Beam Technology •Fundamental Excitatory Modes of the Earth-Ionosphere Cavity

**325 PAGES. 8x10 PAPERBACK. ILLUSTRATED. $16.95. CODE: TTP**

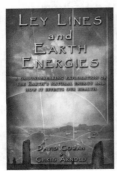

## LEY LINE & EARTH ENERGIES
### An Extraordinary Journey into the Earth's Natural Energy System
### by David Cowan & Chris Arnold

The mysterious standing stones, burial grounds and stone circles that lace Europe, the British Isles and other areas have intrigued scientists, writers, artists and travellers through the centuries. How do ley lines work? How did our ancestors use Earth energy to map their sacred sites and burial grounds? How do ghosts and poltergeists interact with Earth energy? How can Earth spirals and black spots affect our health? This exploration shows how natural forces affect our behavior, how they can be used to enhance our health and well being. A fascinating and visual book about subtle Earth energies and how they affect us and the world around them.

**368 PAGES. 6x9 PAPERBACK. ILLUSTRATED. BIBLIOGRAPHY. INDEX. $18.95. CODE: LLEE**

## THE ORION PROPHECY
### Egyptian and Mayan Prophecies on the Cataclysm of 2012
### by Patrick Geryl and Gino Ratinckx

In the year 2012 the Earth awaits a super catastrophe: its magnetic field will reverse in one go. Phenomenal earthquakes and tidal waves will completely destroy our civilization. These dire predictions stem from the Mayans and Egyptians—descendants of the legendary Atlantis. The Atlanteans were able to exactly predict the previous world-wide flood in 9792 BC. They built tens of thousands of boats and escaped to South America and Egypt. In the year 2012 Venus, Orion and several others stars will take the same 'code-positions' as in 9792 BC!

**324 PAGES. 6x9 PAPERBACK. ILLUSTRATED. $16.95. CODE: ORP**

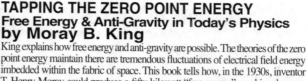

## TAPPING THE ZERO POINT ENERGY
### Free Energy & Anti-Gravity in Today's Physics
## by Moray B. King

King explains how free energy and anti-gravity are possible. The theories of the zero point energy maintain there are tremendous fluctuations of electrical field energy imbedded within the fabric of space. This book tells how, in the 1930s, inventor T. Henry Moray could produce a fifty kilowatt "free energy" machine; how an electrified plasma vortex creates anti-gravity; how the Pons/Fleischmann "cold fusion" experiment could produce tremendous heat without fusion; and how certain experiments might produce a gravitational anomaly.

**180 PAGES. 5x8 PAPERBACK. ILLUSTRATED. $12.95. CODE: TAP**

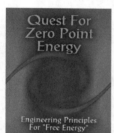

## QUEST FOR ZERO-POINT ENERGY
### Engineering Principles for "Free Energy"
### by Moray B. King

King expands, with diagrams, on how free energy and anti-gravity are possible. The theories of zero point energy maintain there are tremendous fluctuations of electrical field energy embedded within the fabric of space. King explains the following topics: Tapping the Zero-Point Energy as an Energy Source; Fundamentals of a Zero-Point Energy Technology; Vacuum Energy Vortices; The Super Tube; Charge Clusters: The Basis of Zero-Point Energy Inventions; Vortex Filaments, Torsion Fields and the Zero-Point Energy; Transforming the Planet with a Zero-Point Energy Experiment; Dual Vortex Forms: The Key to a Large Zero-Point Energy Coherence. Packed with diagrams, patents and photos. With power shortages now a daily reality in many parts of the world, this book offers a fresh approach very rarely mentioned in the mainstream media.

**224 PAGES. 6x9 PAPERBACK. ILLUSTRATED. $14.95. CODE: QZPE**

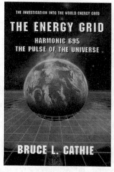

## THE ENERGY GRID
### Harmonic 695, The Pulse of the Universe
### by Captain Bruce Cathie.

This is the breakthrough book that explores the incredible potential of the Energy Grid and the Earth's Unified Field all around us. Cathie's first book, *Harmonic 33*, was published in 1968 when he was a commercial pilot in New Zealand. Since then, Captain Bruce Cathie has been the premier investigator into the amazing potential of the infinite energy that surrounds our planet every microsecond. Cathie investigates the Harmonics of Light and how the Energy Grid is created. In this amazing book are chapters on UFO Propulsion, Nikola Tesla, Unified Equations, the Mysterious Aerials, Pythagoras & the Grid, Nuclear Detonation and the Grid, Maps of the Ancients, an Australian Stonehenge examined, more.

**255 PAGES. 6x9 TRADEPAPER. ILLUSTRATED. $15.95. CODE: TEG**

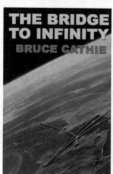

## THE BRIDGE TO INFINITY
### Harmonic 371244
### by Captain Bruce Cathie

Cathie has popularized the concept that the earth is crisscrossed by an electromagnetic grid system that can be used for anti-gravity, free energy, levitation and more. The book includes a new analysis of the harmonic nature of reality, acoustic levitation, pyramid power, harmonic receiver towers and UFO propulsion. It concludes that today's scientists have at their command a fantastic store of knowledge with which to advance the welfare of the human race.

**204 PAGES. 6x9 TRADEPAPER. ILLUSTRATED. $14.95. CODE: BTF**

# ORDER FORM

## 10% Discount When You Order 3 or More Items!

One Adventure Place
P.O. Box 74
Kempton, Illinois 60946
United States of America
Tel.: 815-253-6390 • Fax: 815-253-6300
Email: auphq@frontiernet.net
http://www.adventuresunlimitedpress.com

## ORDERING INSTRUCTIONS

✓ Remit by USD$ Check, Money Order or Credit Card
✓ Visa, Master Card, Discover & AmEx Accepted
✓ Paypal Payments Can Be Made To:
   info@wexclub.com
✓ Prices May Change Without Notice
✓ 10% Discount for 3 or more Items

## SHIPPING CHARGES

### United States

✓ Postal Book Rate { $4.00 First Item
                     50¢ Each Additional Item
✓ POSTAL BOOK RATE Cannot Be Tracked!
✓ Priority Mail { $5.00 First Item
                  $2.00 Each Additional Item
✓ UPS { $6.00 First Item
         $1.50 Each Additional Item
   NOTE: UPS Delivery Available to Mainland USA Only

### Canada

✓ Postal Air Mail { $10.00 First Item
                    $2.50 Each Additional Item
✓ Personal Checks or Bank Drafts MUST BE
   US$ and Drawn on a US Bank
✓ Canadian Postal Money Orders OK
✓ Payment MUST BE US$

### All Other Countries

✓ Sorry, No Surface Delivery!
✓ Postal Air Mail { $16.00 First Item
                    $6.00 Each Additional Item
✓ Checks and Money Orders MUST BE US$
   and Drawn on a US Bank or branch.
✓ Paypal Payments Can Be Made in US$ To:
   info@wexclub.com

## SPECIAL NOTES

✓ RETAILERS: Standard Discounts Available
✓ BACKORDERS: We Backorder all Out-of-
   Stock Items Unless Otherwise Requested
✓ PRO FORMA INVOICES: Available on Request

ORDER ONLINE AT: www.adventuresunlimitedpress.com

---

*Please check:* ☑

☐ This is my first order    ☐ I have ordered before

Name
Address
City
State/Province                          Postal Code
Country
Phone day                    Evening
Fax                Email

| Item Code | Item Description | Qty | Total |
|-----------|------------------|-----|-------|
|           |                  |     |       |
|           |                  |     |       |
|           |                  |     |       |
|           |                  |     |       |
|           |                  |     |       |
|           |                  |     |       |
|           |                  |     |       |
|           |                  |     |       |
|           |                  |     |       |
|           |                  |     |       |
|           |                  |     |       |
|           |                  |     |       |
|           |                  |     |       |

*Please check:* ☑

Subtotal ▶
Less Discount-10% for 3 or more items ▶

☐ Postal-Surface                                Balance ▶
☐ Postal-Air Mail  Illinois Residents 6.25% Sales Tax ▶
   (Priority in USA)                    Previous Credit ▶
☐ UPS                                          Shipping ▶
   (Mainland USA only) Total (check/MO in USD$ only) ▶

☐ Visa/MasterCard/Discover/American Express

Card Number
Expiration Date

## 10% Discount When You Order 3 or More Items!